Mushrooms of Ontario & Eastern Canada

To

Mae

MUSHROOMS of Northeast North America

Midwest to New England

GEORGE BARRON

First printed in 1999 10 9 8 7 6 5

Printed in China

The Publisher: Lone Pine Publishing

10145 – 81 Avenue
Edmonton, AB T6E 1W9 Canada

1808 B Street NW, Suite 140
Auburn, WA 98001 USA

Website: http://www.lonepinepublishing.com

Canadian Cataloguing in Publication Data

Barron, George L., 1928–
Mushrooms of Ontario and Eastern Canada

Includes index.
ISBN 13: 978-155105-199-4
ISBN 10: 1-55105-199-0

1. Mushrooms—Ontario—Identification. 2. Mushrooms—Canada, Eastern—Identification. 3. Fungi—Ontario—Identification. 4. Fungi—Canada, Eastern—Identification. I. Title.

QK617.B37 1999 579.6'09713 C99-910093-9

Editorial Director: Nancy Foulds
Project Editor: Lee Craig
Production Manager: David Dodge
Layout & Production: Gregory Brown
Book Design: Gregory Brown
Cover Design: Rob Weidemann
Cover Photograph: Chanterelle Waxcap by George Barron
Illustrations: George Barron
Cartography: Volker Bodegom
Scanning, Separations & Digital Film: Elite Lithographers, Edmonton, Alberta, Canada

The majority of the photographs in this book are by George Barron. Additional photography is primarily noted in the text and also includes the following photos: Brian Shelton: 7; Greg Thorn: 19 (lower), 320 (upper), 204 (upper).

Warning: Use the utmost **caution** when considering whether or not to eat a mushroom or fungus. If you don't know what kind of mushroom or fungus it is, don't eat it; always remember that eating mushrooms or fungi can be dangerous, and there is **no** guarantee that a mushroom won't cause harm or discomfort. Lone Pine Publishing, its staff and the author bear no liability for the information contained within this book.

The publisher gratefully acknowledges the support of the Department of Canadian Heritage.

PC: P13

contents

LIST OF KEYS AND DIAGRAMS

List of Keys

List of Diagrams

ACKNOWLEDGEMENTS

It is a pleasure to thank Scott Redhead, the acknowledged Canadian authority on gill fungi, for his counsel and advice on the names and distribution of the macrofungi. Scott has guided and helped me for many years through the morass of Hymenomycete taxonomy and for this I am indebted. Scott's enthusiastic companionship in the field, his generous advice and his encouragement and support have been greatly appreciated. Not only that, he's a pretty fair cook!

I also wish to acknowledge the photographers who generously donated their images to help me plug the gaps. I am especially grateful to Greg Thorn and Brian Shelton for their many quality slides. It is a special pleasure to thank the many friends and colleagues for their companionship on forays and field trips over the years and for directing or guiding me to their favourite and treasured spots.

I would like to thank the Kerr Family who sponsored the Kerr Plant Disease and Mycological Herbarium in the Department of Environmental Biology, where much of the work on this book was carried out and where many of the collections are now stored.

I also take this opportunity to thank the Department of Environmental Biology and the University of Guelph for the use of office space and laboratory facilities and for granting the logistical support that made this project possible.

I wish to acknowledge the help of Carla Zelmer, for testing the keys and finding their flaws, and Jenny Shih, for enthusiastic assistance with much of the routine work associated with the preparation of the final manuscript. I must also recognize the contribution of Marc Favreau, whose painstakingly thorough translation of the text into French drew my attention to a number of errors and ambiguities in the science and English.

It is a pleasure to thank Tom Hsiang, whose computer expertise and ready and willing assistance saved me from self-destruction on innumerable occasions.

Finally, I would like to thank Greg Thorn for his critical reading of the manuscript and for the ready and willing expertise that resulted in many useful suggestions and corrections for improvement of the book. I didn't always go along with his recommendations so the mistakes are all mine!

George L. Barron

This reference guide shows a selection of the species included in this book to illustrate their various characteristics.

Slime Moulds

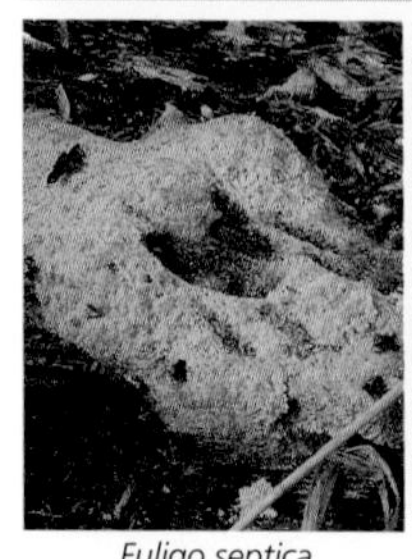

Fuligo septica
p. 36

Diachea leucopodia
p. 38

Arcyria nutans
p. 40

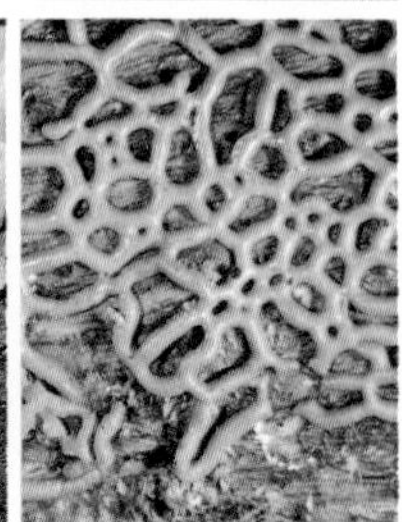

Hemitrichia serpula
p. 41

Metatrichia vesparium
p. 42

Chinese Lantern
Dictydium cancellatum
p. 44

Lycogala flavofuscum
p. 45

Tubifera ferruginosa
p. 46

Sac Fungi

False Orange Peel
Aleuria rhenana
p. 53

Blue-Stain Fungus
Chlorociboria aeruginascens, p. 54

Scarlet Cup
Sarcoscypha austriaca
p. 55

Ascotremella faginea
p. 56

Bulgaria inquinans
p. 57

Lachnellula agassizii
p. 58

Scutellinia setosa
p. 60

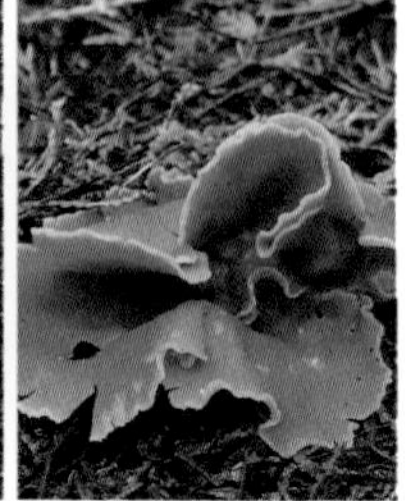

Spreading Cup
Peziza repanda
p. 61

Sac Fungi

Geopora sepulta
p. 63

Devil's Urn
Urnula craterium
p. 64

Ear-Shaped Otidea
Otidea auricula
p. 65

Common Jelly Baby
Leotia lubrica
p. 67

Olive Earth Tongue
Microglossum olivaceum
p. 68

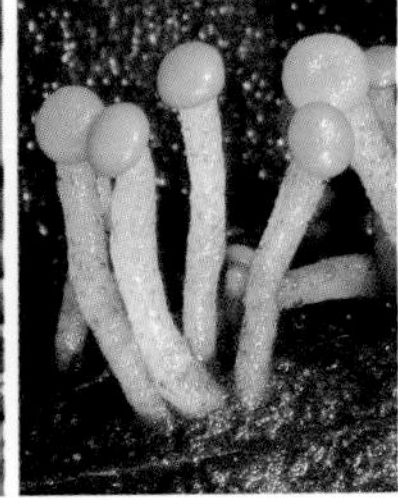
Vibrissea truncorum
p. 70

Velvet-Stalked Fairy Fan
Spathulariopsis velutipes
p. 71

Black Morel
Morchella elata
p. 72

Half-Free Morel
Morchella semilibera
p. 73

White Elfin Saddle
Helvella crispa
p. 75

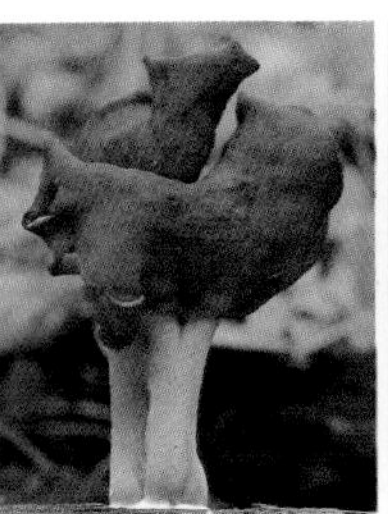
Saddle-Shaped False Morel
Gyromitra infula
p. 76

Helvella macropus
p. 77

Dead Man's Fingers
Xylaria polymorpha
p. 79

Candlesnuff
Xylaria hypoxylon
p. 80

Cordyceps militaris
p. 81

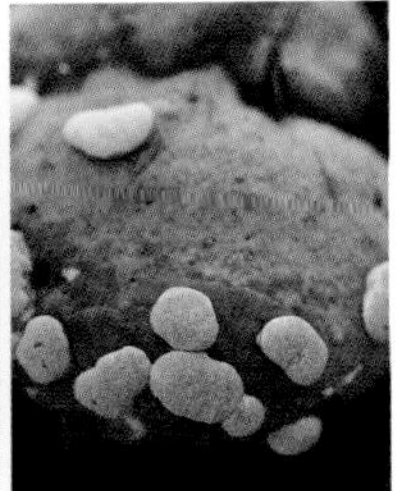
Hypocrea pulvinata
p. 84

Puffballs & Friends

Gem-Studded Puffball
Lycoperdon perlatum
p. 89

Pear-Shaped Puffball
Lycoperdon pyriforme
p. 90

Water Measurer
Astraeus hygrometricus
p. 91

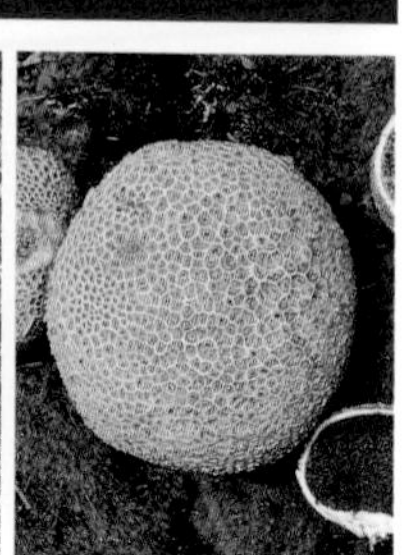
Earthball
Scleroderma citrinum
p. 92

Giant Puffball
Calvatia gigantea
p. 93

Four-Armed Earthstar
Geastrum quadrifidum
p. 95

Fringed Earthstar
Geastrum fimbriatum
p. 95

Beaked Earthstar
Geastrum pectinatum
p. 95

Dung Loving Bird's Nest
Cyathus stercoreus
p. 97

Striate Bird's Nest
Cyathus striatus
p. 98

Grey Bird's Nest
Cyathus olla
p. 98

Sphaerobolus stellatus
p. 98

Ravenel's Stinkhorn
Phallus ravenelii
p. 99

Skirted Stinkhorn
Dictyophora duplicata
p. 99

Dog Stinkhorn
Mutinus ravenelii
p. 100

Lizard's Claw
Lysurus cruciatus
p. 100

Jelly Fungi

Orange Jelly
Dacrymyces palmatus
p. 102

Yellow Staghorn Fungus
Calocera viscosa
p. 103

Leafy Jelly Fungus
Tremella foliacea
p. 104

White Coral Jelly Fungus
Tremella reticulata
p. 104

Black Witch's Butter
Exidia glandulosa
p. 106

Apricot Jelly Fungus
Tremiscus helvelloides
p. 107

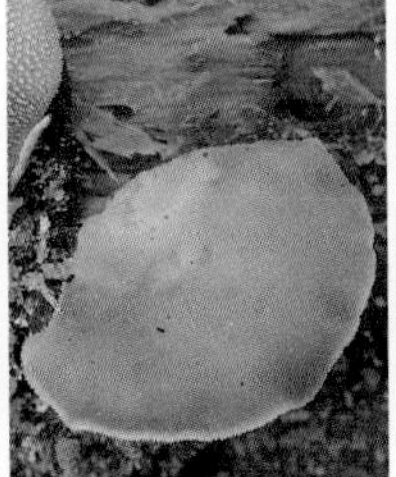

Toothed Jelly Fungus
Pseudohydnum gelatinosum, p. 107

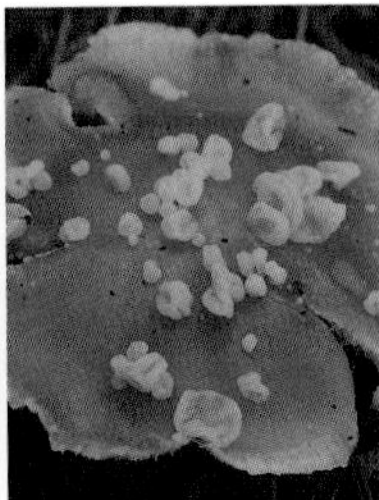

Parasitic Jelly
Syzygospora mycetophila
p. 109

Coral Fungi

Worm-Like Coral
Clavaria vermicularis
p. 111

Rosy Club Coral
Clavaria rosea
p. 111

Flat-Topped Coral
Clavariadelphus truncatus
p. 113

Pestle-Shaped Coral
Clavariadelphus pistillaris
p. 113

Clavulina rugosa
p. 114

Crown Coral
Clavicorona pyxidata
p. 116

Golden Coral
Ramaria aurea
p. 116

Eastern Cauliflower
Sparassis herbstii
p. 120

Tooth Fungi

Comb Tooth
Hericium coralloides
p. 122

Pine Cone Fungus
Auriscalpium vulgare
p. 127

Shelving Tooth
Climacodon septentrionale
p. 128

Steccherinum ochraceum
p. 128

Bracket Fungi

Blue Albatrellus
Albatrellus caeruleoporus
p. 133

Albatrellus confluens
p. 133

Turkey Tail
Trametes versicolor
p. 138

Artist's Conk
Ganoderma applanatum
p. 139

Lacquered Polypore
Ganoderma lucidum
p. 139

Red-Banded Polypore
Fomitopsis pinicola
p. 142

Gloeophyllum sepiarium
p. 142

Oak Polypore
Daedalea quercina
p. 143

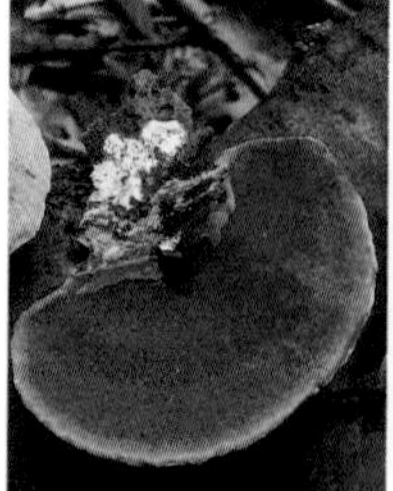
Cinnabar Polypore
Pycnoporus cinnabarinus
p. 149

Carnation Fungus
Thelephora caryophyllea
p. 153

False Turkey Tail
Stereum ostrea
p. 154

Cotylidia diaphana
p. 154

Boletes

Gilled Bolete
Phylloporus rhodoxanthus
p. 159

Old Man of the Woods
Strobilomyces strobilaceus
p. 159

Boletellus
chrysenteroides
p. 160

Parasitic Bolete
Boletus parasiticus
p. 162

Two-Coloured Bolete
Boletus bicolor
p. 163

King Bolete
Boletus edulis
p. 163

Bay-Brown Bolete
Boletus badius
p. 164

Blue-Staining Bolete
Gyroporus cyanescens
p. 165

Orange Bolete
Leccinum aurantiacum
p. 166

Dark-Stalked Bolete
Leccinum atrostipitatum
p. 167

Tylopilus eximius
p. 168

Peppery Bolete
Chalciporus piperatus
p. 170

Slippery Jack
Suillus luteus
p. 170

Painted Bolete
Suillus spraguei
p. 173

Granular-Dotted Bolete
Suillus granulatus
p. 174

Ash Bolete
Gyrodon merulioides
p. 177

Pink-Spored Mushrooms

Salmon-Coloured Nolanea
Nolanea quadrata
p. 184

Green Leptonia
Leptonia incana
p. 185

Pluteus umbrosus
p. 186

Pluteus tomentosulus
p. 188

Pluteus admirabilis
p. 188

The Miller
Clitopilus prunulus
p. 189

Rhodotus palmatus
p. 190

Horn of Plenty
Craterellus fallax
p. 190

Dark-Spored Mushrooms

Agaricus silvicola
p. 192

Shaggy Mane
Coprinus comatus
p. 195

Mica Cap
Coprinus micaceus
p. 197

Coprinus radians
p. 199

Rosy Gomphidius
Gomphidius subroseus
p. 200

Hypholoma capnoides
p. 202

Bell-Shaped Panaeolus
Panaeolus sphinctrinus
p. 205

Psathyrella delineata
p. 206

Brown-Spored Mushrooms

Agrocybe vervacti
p. 212

Dunce Cap
Conocybe lactea
p. 213

Cracked-Top
Agrocybe molesta
p. 214

Flaming Pholiota
Pholiota flammans
p. 216

Pholiota squarrosoides
p. 217

Changeable Pholiota
Pholiota mutabilis
p. 218

Tubaria confragosa
p. 219

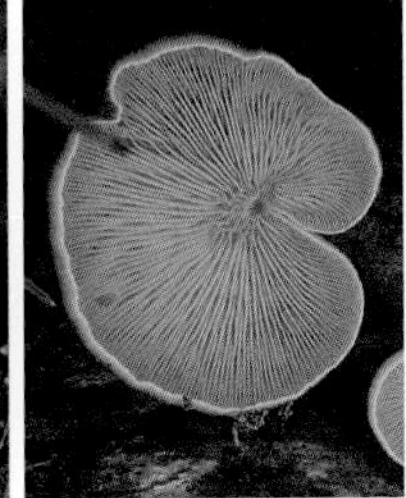
Crepidotus crocophyllus
p. 221

Big Laughing Mushroom
Gymnopilus spectabilis
p. 222

Deadly Galerina
Galerina autumnalis
p. 222

Poisonous Paxillus
Paxillus involutus
p. 224

Cortinarius alboviolaceus
p. 226

Purple Cort
Cortinarius violaceus
p. 228

Deadly Cort
Cortinarius gentilis
p. 229

Cinnamon Cort
Cortinarius cinnamomeus
p. 229

Blood-Red Cort
Cortinarius sanguineus
p. 230

Light-Spored Mushrooms

Yellow Patches
Amanita flavoconia
p. 238

Fly Agaric
Amanita muscaria
p. 239

Gem-Studded Amanita
Amanita gemmata
p. 239

Amanita rhopalopus
p. 240

Yellow Parasol
Leucocoprinus luteus
p. 242

Lepiota cristata
p. 244

Woolly Chanterelle
Gomphus floccosus
p. 249

False Chanterelle
Hygrophoropsis aurantiaca
p. 252

Buttery Collybia
Rhodocollybia butyracea
p. 257

Tuberous Collybia
Collybia tuberosa
p. 259

Marasmiellus candidus
p. 264

Xeromphalina campanella
p. 267

Hygrocybe punicea
p. 269

Hygrocybe vitellina
p. 271

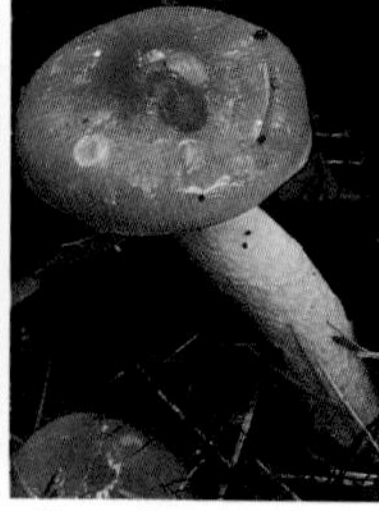

Hygrophorus speciosus
p. 273

Russula Waxcap
Hygrophorus russula
p. 275

Light-Spored Mushrooms

Purple Laccaria
Laccaria amethystea
p. 279

Melanoleuca melaleuca
p. 281

Mycena amabillisima
p. 282

Mycena epipterygia
p. 287

Pleurotus dryinus
p. 292

Late Fall Oyster
Panellus serotinus
p. 293

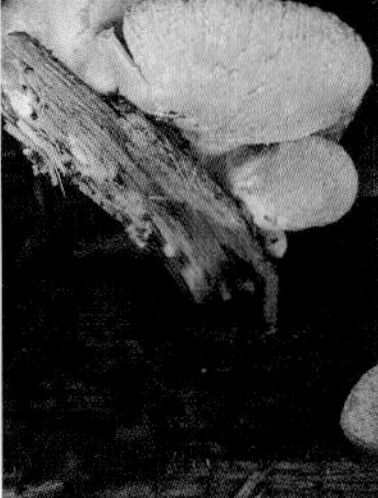
Lentinellus ursinus
p. 293

Jack O'Lantern
Omphalotus olearius
p. 295

Yellow-Green Tricholoma
Tricholoma flavovirens
p. 298

Tricholoma myomyces
p. 299

Megacollybia platyphylla
p. 302

Xerula megalospora
p. 303

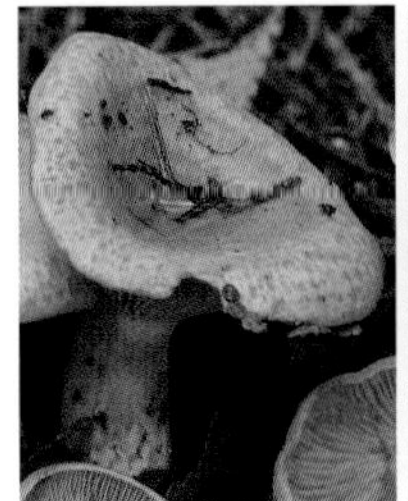
Delicious Lactarius
Lactarius deliciosus
p. 306

Lactarius rufus
p. 308

Russula brevipes
p. 310

Russula decolorans
p. 313

INTRODUCTION

It is estimated that there are more than a million species of fungi in the world. Most of these are microscopic and we are not concerned with them here. Fungi big enough to see with the eye are called **Macrofungi** and there are many thousands of these in Canada, including many that are imperfectly known or have not yet been described.

This book describes and illustrates the macrofungi found in northeastern North America. The territories covered are indicated on the map (p. 21) and range from northern Ontario, Minnesota and Wisconsin in the west to the Maritime provinces and the New England states in the east. All of the states and provinces bordering the Great Lakes and the St. Lawrence River are also served by this guide.

More than 600 species of fungi are described and I have favoured those that are widespread, common or of striking appearance. As much as possible, I have also given a balanced treatment to the many different groups of fungi represented across the eastern regions. Some species covered in this book are found only in the coastal provinces and states; other species are found only in the Great Lakes region. Generally, however, fungi are cosmopolitan in their distribution and a large majority of the species described fruit across the entire eastern zone.

The 609 species in this guide can be separated into nine groups. The groups recognized here (see picture key, pp. 30–31) are called Slime Moulds, Sac Fungi, Puffballs and Friends, Bracket Fungi, Jelly Fungi, Coral Fungi, Tooth Fungi, Boletes and Gill Fungi. I have also discussed miscellaneous fungi in sidebars scattered throughout the introduction and appendix. These fungi are found around the home and garden; they are very common but do not fit happily into any of the other groups.

How to Use This Book

To identify a fungus, it is first necessary to place it in the correct group. To follow this step, use the "**picture key**" (pp. 30–31) to groups of fungi. The range of fungi found within each of the nine groups is represented by four pictures. Each group has a common name that emphasizes a key characteristic that helps in identification, e.g., puffballs, jelly fungi, tooth fungi. With a little practice you will become familiar with these features and be able to identify a

Russula decolorans, a gilled fungus and light-spored mushroom.

fungus to group without difficulty. When a fungus has been identified to group, the picture key indicates the page numbers in the book where the illustrations and descriptions for species in that group are located. An introductory section at the beginning of each group gives more details that help to confirm your decision.

Identification to Genus

Once the specimen is placed in the correct group, it can be identified to genus and species. A popular method is to leaf through the illustrations of the group searching for a fungus similar to the one you have found. Alternatively, you can use a "**Key to Genera**" when available. Leafing through the pages works well with small groups such as Slime Moulds (28 species), Jelly Fungi (18 species), Tooth Fungi (16 species) and a few others. In the case of Sac Fungi (78 species), Bracket Fungi (61 species) and Gill Fungi (about 300 species), however, the groups are too large to be handled efficiently in this manner. For these larger groups, therefore, I have constructed **dichotomous keys**. There is an old adage that says, Keys are constructed by people who don't need them for people who can't use them. There is a lot of truth to this statement, but don't be discouraged by early failures. Persistence with the keys will pay off in the long run when you advance to the next level of mushrooming and use larger books with more detailed keys.

Beauveria bassiana

Beauveria bassiana is one of the most successful of the fungal parasites of insects. It attacks a wide range of insect hosts and has been used on several occasions to effect biological control of various insect pests. In the above photo, we see a cicada—the insect that sounds like a buzz saw and drives some people crazy—which has been killed by *Beauveria bassiana*. The fungus proliferates inside the body of the host and then bursts out between the body segments where it produces powdery, white spore-masses.

The dichotomous key is so-called because it is based on a simple choice of two options. You select the choice that fits best, which leads you to the next appropriate number in the key and eventually to a **genus** name. In the dichotomous key one or two critically important features of a fungus are stressed to the exclusion of all others. A wrong decision, therefore, can easily lead you far astray, demonstrating that dichotomous keys have severe limitations. In the not-too-distant future, we will have computerized programs where we can just plug in all the features of the fungus we find, press the ENTER key and up will pop a genus and/or species name. In field guides such as this one, however, we might have to stick with the old ways for a while longer.

It is good idea to practise using a key with a fungus you already know; use the key to confirm its identity. If you go wrong you can see how and perhaps why and make appropriate corrections in your thinking. With practice you will be able to key out the "unknowns" with more confidence. Don't be discouraged by a lack of success. Often, the keys themselves are at fault. One reason is that distinctions between genera are sometimes based on microscopic features that are not applicable to a field key and secondary features must be chosen that are not as discriminating. Sometimes the specimens are too old or too young or perhaps are not "typical" of the species.

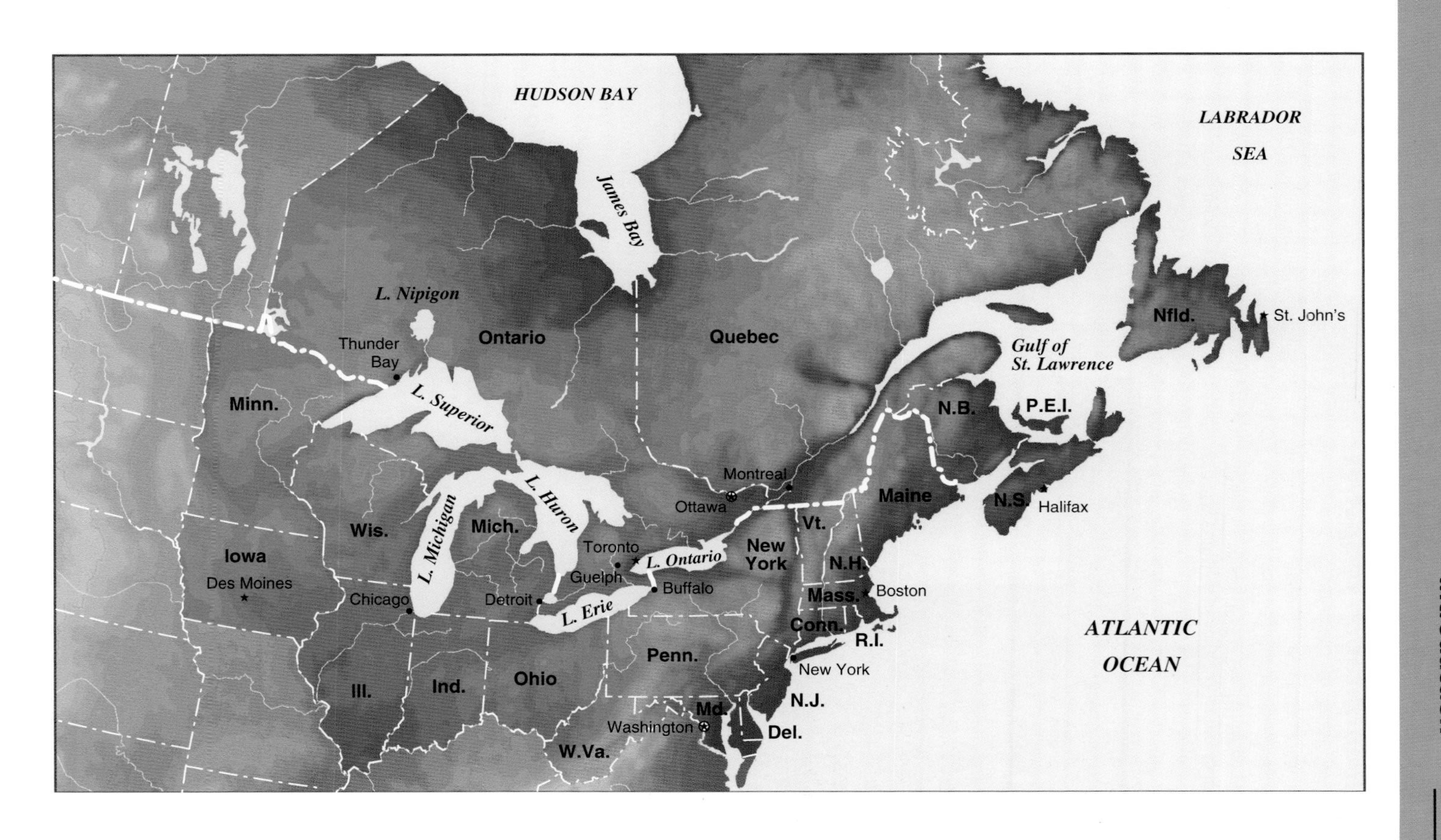
HUDSON BAY
LABRADOR SEA
James Bay
L. Nipigon
Thunder Bay
Ontario
Quebec
Nfld.
St. John's
Gulf of St. Lawrence
Minn.
L. Superior
N.B.
P.E.I.
Montreal
Ottawa
Maine
N.S.
Halifax
L. Huron
Wis.
L. Michigan
Mich.
Vt.
Iowa
Des Moines
Toronto
L. Ontario
New York
N.H.
Guelph
Buffalo
Mass.
Boston
Chicago
Detroit
L. Erie
Conn.
R.I.
ATLANTIC OCEAN
Penn.
New York
Ill.
Ind.
Ohio
N.J.
Md.
Washington
Del.
W.Va.

Identification to Species

The **Key to Genera** will guide you to the appropriate section in this book that covers that genus, and you can compare your specimen with the photographs of the species available. The large colour photographs have been selected to show specimens that are "typical" of the species and for the most part are mature specimens in prime condition. The descriptive paragraph that accompanies each photograph gives more details and expands the information to cover a wider range of colours, sizes, habitats, etc., than are indicated in the photograph. As fungi age, features important in identification can change considerably: colours might fade, rings on the stem might disappear, patches on the caps might wash off, gelatinous fungi might dry up, gills might discolour, etc. Size ranges must not be interpreted too literally. It is not unusual to find mushrooms well outside the ranges given. I have found a tiny, perfectly formed *Chalciporus piperatus* that could fit comfortably on a dime (would I had taken the picture), but I did not extend the range of sizes for this species because the specimen was too far outside the "norm." Colours are also very subjective and must be interpreted with a fair amount of latitude.

Mycetophilids

Some insects, such as many sciarids and phorids, enjoy eating mushroom fruitbodies even more than we do, and they depend on fungal fruitbodies for their survival. These fungus-loving insects are called mycetophilids. They lay their eggs in the flesh of a mushroom. The eggs hatch out and the larvae consume the mushroom. Fungi have various ways to protect themselves from insect attack: they can produce lethal or noxious chemicals; sometimes they have hairy surfaces that make it difficult for the insect to lay its eggs; or they produce slimy goo that traps the insect. Here (above photo), we see a mycetophilid that has come to an untimely end after becoming trapped in the thick slime, which envelopes the stalk of *Mycena rorida*.

Species Descriptions

Each species description outlines features important for identification. In many cases, the fungi are so distinctive that they cannot be mistaken and the description can be circumscribed in a few phrases. In other cases, there might be a number of similar species, making it more difficult to identify the specimen with certainty. Remember, there are many thousands of macrofungi in North America and most of them are not found in this, or any other, guide. Very often, therefore, you must be content to identify a specimen to the genus only and cannot worry about the identity of the species. Don't feel badly about this; even the most experienced amateurs or professional collectors have the same problem. For example, there are more than 600 species in *Cortinarius* and only a dozen are described in this book (pp. 226–30). *Cortinarius* species are very common so you are going to find many that you cannot name to species. Identifying a mushroom collection as a "Cort" is very acceptable.

Scientific Names and Common Names

In the species descriptions there is always a scientific name beside or below the photograph in the colour bar and, for some of the species, there is also a common name below the colour bar. Common names are usually restricted to fungi that, for one reason or another, are interesting, important or very common. Unfortunately, common names change from place to place and consequently can be different in different books. Some books have translated the scientific name as a common name or created common names *de novo*, but this approach is not particularly advantageous. In the long haul, for better or worse, the scientific binomial is the name of choice.

One of the most frustrating aspects of mycology (the study of fungi) is that the scientific names constantly change, because mycology is a relatively young science. Gill fungi, in particular, have this problem, and many in this guide have a different name than the one you will find for the same species in other guides. Not all of these changes will stand the test of time so the best temporary solution is to "make haste slowly" when it comes to learning new names. Because the scientific names keep changing, I have often given an alternate name at the end of the descriptive paragraph.

Spinellus fusiger Pin Mould of *Mycena*

Mycena is a successful genus of mushrooms, with over 250 species in North America. The mushrooms commonly grow on woody debris or sometimes in troops of hundreds on the forest floor. During prolonged wet periods, the fruitbodies of *Mycena* are attacked by the species *Spinellus fusiger*, which then produces tall, elegant fruiting stalks radiating from the cap. Each stalk terminates in a tiny, spherical spore ball, giving it a "pin-head" appearance. Infected *Mycena* caps have a striking, hairy appearance.

To Eat or Not To Eat

With all of the uncertainty associated with mushroom identification, mistakes are easily made. **It is clear, therefore, that unfamiliar mushrooms must never be eaten without first having the edibility of the mushroom confirmed by someone with an authoritative knowledge.** Also, remember the personal factor—some people can eat and enjoy mushrooms that might cause others severe alimentary upsets. So the advice of "a friend who has eaten this species often" is not always valid. Nevertheless, there are some mushrooms that are distinctive in appearance, difficult to confuse with poisonous species, and are relatively tasty (each to his own!). There is an illustrated list of these readily identifiable and recommended edible species on pp. 317–18.

Please note that although a mushroom's taste is sometimes described among its identifying features, that is **not** an invitation to eat the mushroom—it still might be poisonous. If you are brave enough to find out what an unidentified mushroom tastes like, chew a tiny piece (measured in cubic millimetres) briefly and then spit it out; do not swallow it.

Gill Fungi and Non-Gill Fungi

Traditionally, mycologists have divided fungi into Non-Gill Fungi and Gill Fungi. A distinguished mycologist, Alexander Smith, of the University of Michigan, published books on fungi under these headings many years ago, and this division has proved to be a very useful method for separation. The separation is for convenience only and is not a division into natural groupings. The Non-Gill Fungi include Slime Moulds (Division Myxomycota), Sac Fungi (Division Ascomycota) and all species of the Division Basidiomycota without gills, i.e., Puffballs, Jelly Fungi, Coral Fungi, Tooth Fungi, Bracket Fungi and Boletes. The Gill Fungi include only those species of Basidiomycota with gills. In this book the Gill Fungi are approximately equal to the Non-Gill Fungi in the number of species treated. Thus, when identifying fungi, it is easy to make a decision that will remove half the species in the book from consideration.

Gill Fungi is by far the largest group of macrofungi, and in this field guide I describe nearly 300 species. For convenience Gill Fungi is broken up into four subgroups based on the colour of the spores. This feature is determined by means of a **spore print** (see p. 156). The subgroups of Gill Fungi are Pink-Spored, Dark-Spored, Brown-Spored and Light-Spored. Recent studies, however, have shown that spore colour is not always a good criterion for indicating natural relationships and other criteria, e.g., morphology and anatomy, have taken precedence over spore colour in the classification of Gill Fungi. While arrangements based on spore colour are breaking down, this method is still the most convenient way to organize Gill Fungi and is the one followed here.

Frost's Amanita (*Amanita frostiana*) is a member of the Division Basidiomycota.

Classification of Fungi

The Kingdom **FUNGI** is divided up into a number of major groups, each called a **division**. Divisions end in the suffix *-mycota*. The principle divisions treated in this book are **Myxomycota** (Slime Moulds), **Ascomycota** (Sac Fungi) and **Basidiomycota** (mushrooms and their relatives). These three divisions have a different "look" and with a little practice, or by leafing casually through this book from time to time, you will find it fairly easy to fit each fungus that you discover into its proper division. Divisions are broken down into smaller groups called **classes**. Class names end with the suffix *-mycetes*. For example, the Division Basidiomycota includes the Classes **Hymenomycetes** and **Gasteromycetes**. The Class **Hymenomycetes** includes the groups Coral Fungi, Bracket Fungi, Boletes, Gill Fungi, etc. The Class **Gasteromycetes** (Puffballs and Friends) includes puffballs, earthstars, stinkhorns and bird's nest fungi. Classes are broken down further into orders and families, but we are not concerned with these finer levels of classification here and have used the common group names instead.

Entomophthora The Fly Killer

In warm days of spring, flies often emerge by the dozens and buzz around the windows. These flies are cluster flies that parasitize earthworms. They overwinter inside the walls of houses and emerge on warm, humid days to lay their eggs on the worms. Occasionally, you will find a fly stuck to the window by the proboscis, with a white halo encircling the body. The fly has been killed by the fungus *Entomophthora*—the halo is caused by masses of spores that have been shot off from the body. *Entomophthora* attacks many different fly species. Here (in photo), an adult of the root maggot (*Delia*) has been attacked by *Entomophthora*. The fungus is seen bursting out between the abdominal segments as a spongy mass.

For individual species of fungi, we use the **binomial** (Linnean) system where each organism has a **genus** name and a **species** name, e.g., *Geastrum saccatum* (p. 96). Where possible, the genus name is based on a feature that is common to all species of the genus. For example, there are a number of species in the puffball group in which the outer wall splits and the segments reflex to form a star-like base. Because of this similarity in structure, almost all of these species are placed in the same genus, *Geastrum*. The genus name Geastrum has its roots in *geos* (earth) and *aster* (star). The species name is like an adjective and usually (but not always) describes some important or interesting feature peculiar to that species. For *Geastrum saccatum*, the species (descriptive) name saccatum indicates that in this earthstar the spore sac nests in the reflexed arms. There are exceptions to every generalization and not all star-shaped fungi are placed in *Geastrum*. The Water Measurer, *Astraeus hygrometricus*, is earthstar-like, but in this species the arms open and close in response to wetting and drying (see p. 91).

Astraeus hygrometricus

Life Cycle of a Mushroom

Under suitable conditions of moisture and nutrition, the mushroom spore germinates to produce a branching network of cylindrical fungal threads, each called a hypha. The hyphae permeate the soil or woody substrate and absorb nutrients through their walls. When a hypha meets another of the same species, but of opposite mating type, the two fuse. New hyphae form an extensive hyphal system (hundreds of kilometres) that colonizes the substrate. Sooner or later, when moisture conditions are just right, the mushroom hyphae will produce a fruitbody, in this case a mushroom. The gill plates hang down from the underside of the mushroom's cap. The spore mother cells (basidia) are produced in fertile layers (hymenium), which cover the surface of the gills. The spores are then shot off and are carried by the wind to new sites to begin a new cycle of growth and exploitation.

Role of Fungi in the Ecosystem

Fungi play many significant and varied roles in our lives. Yeast fungi are responsible for the leavening of bread and for the production of wines and spirits. Fungi are an important source of pharmaceutical compounds such as antibiotics (e.g., penicillin, cephalosporin, griseofulvin). The great resurgence in human organ transplants is thanks to the anti-rejection drug cyclosporin, which is produced by a microscopic fungus called *Tolyplocladium*. Billions of dollars worth of crop losses are caused annually by fungal diseases of plants. Fungi belonging to the

Fig. 1. **Wood Decay.** Fungi are the best wood rotters in the natural world. Breakdown of lignin and cellulose is a prime role of these fungi, and they return staggering levels of carbon dioxide to the atmosphere annually (measured in hundreds of billions of tonnes). The inset photo shows a stump that after several years has become colonized by *Ganoderma applanatum*, one of the pioneer wood decay fungi. After many years and a succession of fungi, insects, bacteria and a variety of microscopic life forms, logs or stumps are reduced to dust. Many wood-rotting fungi attack the wood at lines of weakness, such as vascular rays and the junctions of annual rings. As it dries out, the wood cleaves along these lines. In a large tree a small section of the circumference gives the illusion of a straight line and the pieces of wood appear to have broken at right angles giving a cubic appearance. This type of decay is called **Cubic Rot** (see top photo). It takes 20 years or more to reduce a hardwood trunk to dust.

ringworm group are common causes of skin infections in humans, and there are many other fungi pathogenic on humans and animals. These diverse roles, however, are trivial compared with the impact fungi have in the natural environment and particularly in the forest ecosystem.

Biomass of Fungi

Biomass is the amount of living stuff in a particular system. In forest soils 90 percent of this living stuff (excluding plant roots) is fungal. The other 10 percent is made up of bacteria, protozoa, nematodes, rotifers, springtails, insects, insect larvae, annelids (including earthworms), algae, etc. In other words, actively growing fungi outweigh all other groups combined by a factor of almost 10 to 1! This living fungal material includes two major components and a number of minor components. The two major biomass components in forest soils are mycorrhizal fungi and wood decay fungi.

Mycorrhizae—The Fungus Roots of Trees

Many people believe that forest trees take up water and nutrients through root hairs. This belief is quite wrong. Forest trees are dependent for their survival on the fungi that are associated with their roots. Each tree has hundreds of thousands of kilometres of fungal threads (hyphae) associated with its roots (see "Life Cycle of a Mushroom," p. 26). It is these mycorrhizal fungi that supply the tree with the nutrients and water essential for healthy growth. In exchange, the tree gives the fungus sugars that are manufactured through photosynthesis in its leaves. Using this sugar for energy, the fungus maintains a hyphal grid that permeates the soil to supply the tree. Some of the tree's sugars are also used to produce the fungal fruitbodies. Many of the fungi in this book are mycorrhizal with certain forest trees.

Biodegradation—Fungi and the Carbon Cycle

Fungi are the best wood destroyers. They have the unique ability to physically penetrate the hardest wood and then enzymatically digest its constituents (lignin, cellulose and hemicellulose). Biodegradation of woody stuff returns 80 billion tonnes of carbon to the atmosphere each year. Most of the carbon is produced as carbon dioxide by the fungi that play the major role in the carbon cycle. Amongst the fungi, the Basidiomycota and the Ascomycota contain the major wood rotters. Many of the fungi illustrated in this book are major causes of wood rot.

Wood is made up principally of cellulose and lignin. Some fungi only attack cellulose. Cellulase enzymes, produced by these fungi, remove the cellulose and leave the lignin behind untouched. The lignin that remains gives the wood a brown colour, and this type of decay is called "Brown Rot." Some fungi attack both cellulose and lignin and when these are destroyed the residue has a white or bleached appearance. This type of rot is called "White Rot."

When brown-rot fungi attack wood, they follow the lines of least resistance. The fungi attack the soft vascular rays that run radially across the wood. They also follow the lines of weakness between annual rings of growth. When the wood dries out, it fractures along these primary attack sites and breaks up into angular pieces. This type of decay is referred to as "Cubic Rot" (Fig. 1) and is produced by many species of brown-rot fungi.

Photographing Fungi

Arcyria denudata

Photographing fungi is a satisfying hobby, and it is surprising how few people have taken the trouble to learn the techniques. It is more fun photographing fungi than eating them (see appendix, p. 317) and less risky. Any good single lens reflex camera will do, from a Pentax K to a state-of-the-art Nikon F5. In most situations you can do as well with the cheaper basic camera as an expensive sophisticated one, although dedicated flash systems and metering choices on the more expensive cameras increase your range of opportunities.

You can use close-up attachments such as portrait lenses (1–3 diopter) or a zoom lens with close-up capability, but a true macro lens (fixed focal length) makes it a lot easier to get quality results. With the current emphasis on autofocus lenses, a used manual macro lens in excellent condition can be bought for less than half the original cost. If you only have one macro lens, then choose a focal length of 50–60 mm. If you can afford a second macro, then the 100–105 mm range allows you to get further away from the subject for a better angle of view.

Pholiota flammans

It is important to have a "ground" tripod, where the centre post unscrews and the legs "break" so you can lower the head close to ground level. A tripod with a rapid release gizmo is a real advantage because you can set up quickly. Never wander from site to site with the camera still attached to the tripod. In the woods, exposures of several seconds are common, so you must use a cable release. A couple of reflectors help to brighten the scene and/or lighten up the shadow areas. Reflectors are made from 8" X 10" pieces of cardboard (e.g., old mounting board) covered with aluminum foil. A bean bag, to get an ant's eye view of things, is an optional extra. Use the film of your choice; Kodachrome 64 or Fujichrome 100 are both good choices.

Camarophyllus pratensis

Bondarzewia berkeleyi

Once you find a nice group of fungi, check different angles and heights for the best composition. A common error is to have too much background and the subject mushroom is too small. Try to fill the frame. Most cameras give you a little more coverage than you see in the viewfinder. Nikon viewfinders show you 100 percent. Use the reflectors to add some light to the shadow areas under the cap or brighten the foreground. You can prop up the reflectors with a stick or a metal tent peg to get the right angle of reflectance. If you have some willing help, a little sunlight can be reflected from 3 metres (10 feet) or more away. For good detail you want all the depth of field the conditions will allow, so aim for an aperture of f11–f22 for maximum depth. In the beginning, bracket your shots until you find, with experience, the proper exposure for your system. In deep woods, the exposures are often measured in seconds, and with long exposures you get reciprocity failure that affects both the colour balance and the speed of the film. Keep records of what you're doing so that, when you get the pictures back, you can make a critical evaluation of your results and know what corrections to make the next time. Best of luck!

Erysiphe chichoracearum (and friends)
Powdery Mildews of Ornamentals

The powdery mildews are a remarkably successful group of parasites that attack tens of thousands of different plants. Hosts include shrubby plants, such as lilacs and roses, and annual or perennial flowers, such as phlox, zinnias and begonias. The symptoms (as shown at right) are a white, powdery growth over the infected parts of the plant. These growths are the summer spores. Later in the year, you will see tiny, black dots against the white. These black dots are the sexual fruitbodies of the fungus that allow the fungus to overwinter and attack next year's crop.
Burn or compost thoroughly all the old, infected leaves to prevent carry-over of the fungus.

MYXOMYCOTA

Slime Moulds pp. 33–46

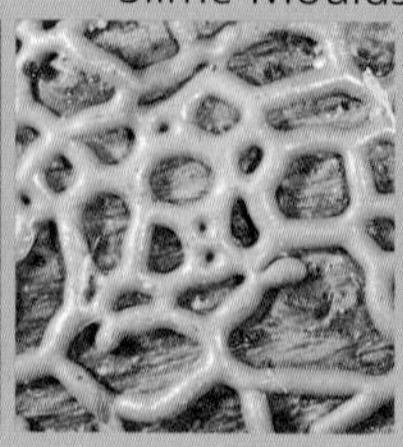

ASCOMYCOTA

Sac Fungi pp. 47–84

BASIDIOMYCOTA

CLASS GASTEROMYCETES

Puffballs & Friends pp. 86–100

CLASS PHRAGMOBASIDIOMYCETES

Jelly Fungi pp. 101–09

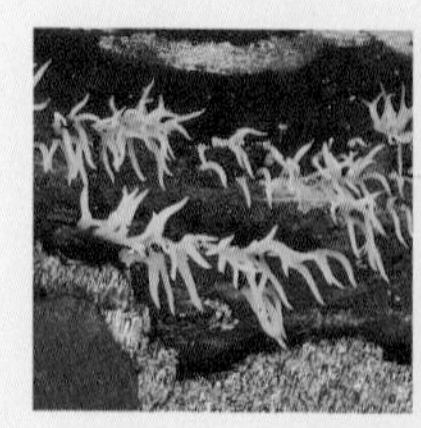

BASIDIOMYCOTA

CLASS HYMENOMYCETES

Coral Fungi pp. 110–20

Tooth Fungi pp. 121–28

Bracket Fungi pp. 129–55

Boletes (Sponge Mushrooms) pp. 156–77

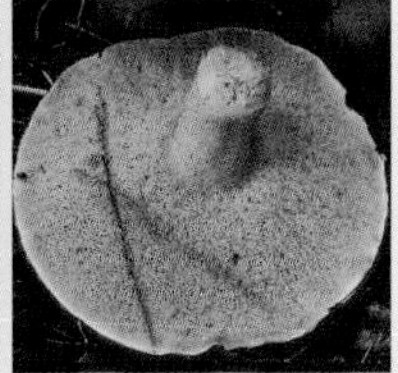

Gill Fungi pp. 179–315

NON-GILL FUNGI

SLIME MOULDS

Myxomycota

Sometimes, after soaking summer rains, you can find a frothy yellow, white or pink slime on a log or stump or even smothering a plant on the forest floor. This strange object is a **Slime Mould**, composed mainly of a slimy mass of protoplasm called a **plasmodium** (see Fig. 2). Most of their lives, slime moulds are hidden inside well-rotted logs or stumps, or buried in leaf mould. When it's time to fruit, however, they migrate to a better site for spore dispersal. The plasmodium can travel several feet and climbs any object, living or dead, that gives it a site advantage. So, it's not unusual to find slime moulds fruiting on green plants, dead twigs, old polypores, stumps, logs or at eye level on a living tree trunk!

Fig. 2. **Slime Mould Plasmodium**. The plasmodium (slime stage) of this slime mould is travelling over the log to a suitable site for fruiting. It moves in veins and can travel several metres. It is very sensitive to drying, so usually travels during the night, and must get to where it is going as quickly as possible before it dries out.

Slime moulds don't fit easily into our classification system: they move and feed like animals; they engulf all kinds of organic particles in their paths, like giant amoebae; they digest what they can; and, animal-like, they violently eject unwanted particles. Slime moulds are fungus-like, however, in producing tiny fruitbodies that contain spores, which are dispersed by wind. There are about 500 known species, and many of these have a global distribution, indicating that slime moulds have been around for a long time.

The slime stage doesn't help much in identifying species of slime moulds, because they often look much alike at this phase. When they fruit, however, we can separate species by the shape, size and colour of the fruitbody and spores. Many of the common species can be identified using a simple 10X hand lens (loupe). For serious students of the group, however, microscopic characteristics are important. For those who have a microscope, a useful book called *How to Know the True Slime Moulds* by Marie Farr gives keys, descriptions and illustrations for most of the slime moulds in North America.

There are four types of fruitbodies in slime moulds. The commonest is the **sporangium**. Here, all the protoplasm separates into pieces and each piece forms a tiny fruitbody. Sporangia are usually only a few millimetres tall, but in some exceptional species, such as *Stemonitis axifera* (p. 43), they might reach several centimetres in height. Hundreds or even thousands of sporangia, more or less the same size and shape, can be produced simultaneously from a single plasmodium.

A relatively large fruitbody, which is very variable in size, is the **aethalium**. To form an aethalium, all of the protoplasm collects into one large piece or sometimes into several pieces of different shapes and sizes, and these mature to form one or several fruitbodies. Some fruitbodies, such as *Fuligo septica* (p. 36), form crusty masses more than 15 cm long. Because of their large size, only one or a few aethalia are produced at a time. In *Lycogala* spp. (p. 45), the aethalia are smaller, usually 1 cm or less, and are sometimes produced in large numbers. In such cases the aethalia might be mistaken for sporangia, but there is a large variability in the size of the fruitbodies.

The third type of fruitbody is called a **pseudoaethalium**. Here, what appears to be an aethalium is actually a large number of individual sporangia that appear to be fused together. This fusion forms the pseudoaethalium, as in *Tubifera ferruginosa* (p. 46). The prefix *pseudo-* is often used in mycology and means "false."

A network of fat veins over the surface of a log or stump is the **plasmodiocarp**, the last type of fruitbody. To form a plasmodiocarp, all the protoplasm moves to the central veins, which then mature and differentiate to form the fruitbody. *Hemitrichia serpula* (p. 41) is the most common species of slime mould producing plasmodiocarps.

Note: For fungi in general, and slime moulds in particular, the records of distribution are poor or nonexistent. For example, from the scant records available, it is known that a slime mould, such as *Physarella oblonga* (p. 36), is widespread, but not whether it is common or rare. Slime moulds, in general, have a wide distribution.

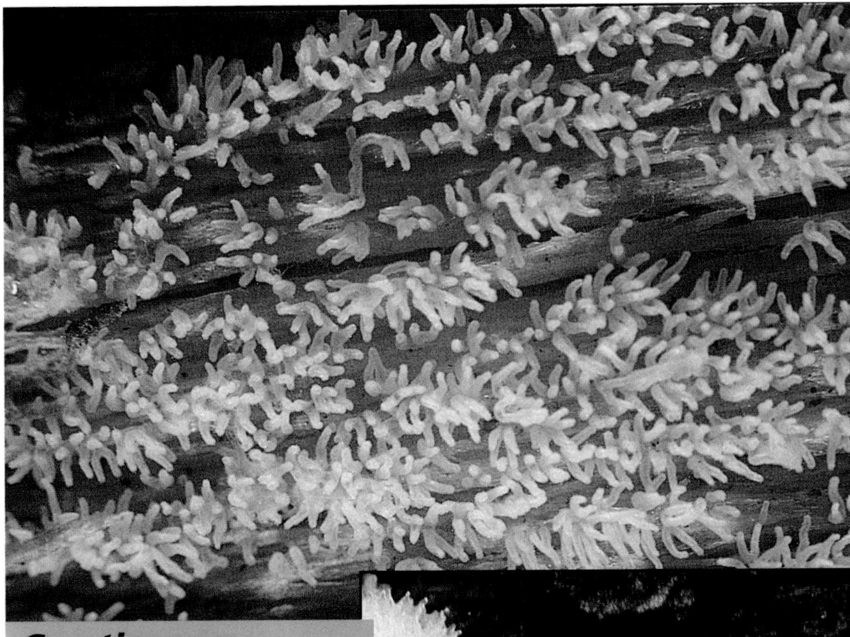

Ceratiomyxa fruticulosa

This species has two varieties that are illustrated here. Fruitbodies of *C. fruticulosa* var. *fruticulosa* (top) form tiny, white columns only a few millimetres tall. Under a hand lens, they have a fuzzy appearance, because of the production of spores on the surface. Fruitbodies are very delicate and, as for many slime moulds, can be obliterated by the touch of a finger. Fruitbodies of *C. fruticulosa* var. *porioides* (right) are dome-shaped, white, honeycombed, and up to 1 cm across. Widespread and common, this slime mould fruits in scattered clusters over well-rotted logs. Both varieties are common after prolonged rainy periods.

Fuligo septica

Fruitbody (aethalium) is a cake-like mass, up to 20 cm in the longest dimension by 3 cm thick, white, yellowish, ochre, or red-brown, and with a smooth but brittle crust, which breaks away to reveal a black spore-mass. Widespread and common, *F. septica* can migrate 1 m or more to fruit on stumps, logs, or living plants, often in the rich soil of well-manured gardens.

Mucilago crustacea

Fruitbody (aethalium) is a cake-like mass, up to 7 cm long by 5 cm wide and 2 cm thick. Outside wall is composed of a layer of crystalline, chalky material that gives it a white, crusty texture. Spore-mass inside is black. This species fruits on decaying leaves and rotting wood.

Physarella oblonga

Fruitbodies (sporangia) are up to 3 mm tall, with nodding, cylindrical heads that split into lobes, which reflex to expose a yellow, trumpet-shaped corona and spiny threads. Stalk is reddish. The mature fruitbody looks like a miniature daffodil. This pecies fruits on dead wood and leaves.

Leocarpus fragilis

Fruitbodies (sporangia) are 2–4 mm tall, ellipsoid to egg-shaped, smooth, shining, brittle, stalked, hanging, and ochre to chestnut-brown at maturity. Spore-mass is black, with white flecks. Widespread and common, this species fruits on leaves, twigs, living or dead plants, etc.

This specimen has climbed up the stem of a plant and produced egg-shell fruitbodies, which hang in clusters from the stem. During its journey, part of the plasmodium has crossed over to a grass blade that is touching the stem of the plant. Climbing the plant gives the slime mould a better site for spore dispersal.

Diachea leucopodia

Fruitbodies (sporangia) are 1–2 mm tall, with a chalk-white stalk and a short-cylindrical, iridescent, purple-black head. Sporangia arise in large numbers from a white crust. Black spore-mass is supported by a stout, white column through the centre of the head. Widespread and not uncommon, *D. leucopodia* fruits on leaves, sticks, and living plants.

Badhamia utricularis

Fruitbodies (sporangia) are globose to pear-shaped, about 1 mm across, blue-grey to grey-black and iridescent, and hang in clusters from yellowish strands. Striking in the early stages when the developing sporangia are a brilliant orange-yellow. Spore-mass is black and flecked with chalky, white threads. Widespread and common, this species fruits on rotting wood.

Photo (left): Greg Thorn.

Physarum cinereum

Fruitbodies (sporangia) are blue-grey, up to 1 mm thick by 0.5 mm wide by several millimetres long, crowded and sometimes fused, and have no stalk (sessile). Spore-mass is purple-black, with white flecks. This species is widespread and common, and fruits on leaf litter, strawberry plants, etc. It is often found on lawns where fruitbodies might cover several square feet of grass. It does no harm and can be washed off with a hose or power mowed!

Hemitrichia species

Young Sporangia

Young fruitbodies (sporangia) of *Hemitrichia* attract the eye by their brilliant red to orange colour, which stands out in sharp contrast to the woody substrate or adjacent plants. At this stage sporangia are very soft and will smear if touched. They cannot be identified until they mature and turn mustard-yellow (p. 41).

Arcyria denudata

Fruitbodies (sporangia) are up to 3 mm tall, egg-shaped to short-cylindrical, and brick-red to purple-brown, sometimes fading with age to a drab brown. Outer wall disappears, and the contents expand to form a bright purple-red, cottony mesh, also fading with age. This species is one of the most common of the slime moulds and produces extensive fruitings of hundreds of sporangia on well-rotted logs.

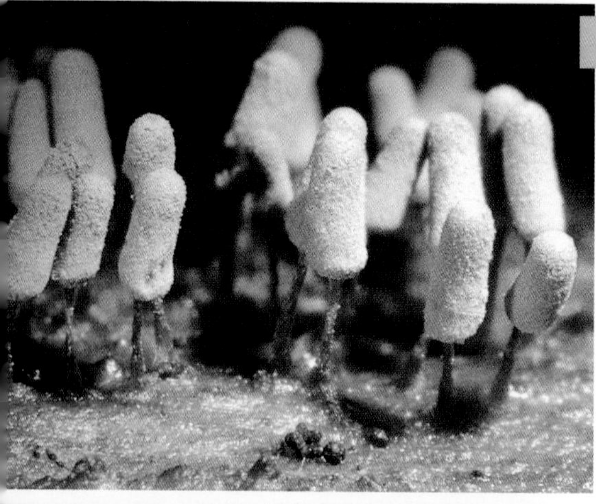

Arcyria cinerea

Fruitbodies (sporangia) are short-cylindrical, pale grey, with blackish stalks, and up to 4 mm tall. Sometimes 3–7 individual sporangia are clustered on a common stalk. *A. cinerea* is one of the most common of the slime moulds in our region. It fruits on rotten wood in scattered troops, sometimes involving hundreds of sporangia.

Arcyria nutans

Fruitbodies (sporangia) are cylindrical and about 2 mm tall, and have no stalk (sessile). Outer wall disappears, however, and the contents expand to form a floppy, cylindrical, yellow to ochre spore-head, up to 12 mm long. *A. nutans* is widespread and not uncommon. It fruits on well-rotted wood in scattered clusters.

Hemitrichia calyculata

Fruitbodies (sporangia) are 1–3 mm tall, top-shaped, mustard-yellow to olive-yellow, and with a dark red-brown stalk. Outer wall disappears, and the contents expand to form a clavate, cottony head supported by a shallow, membranous cup (calyculus). This species is one of our most common slime moulds and fruits on well-rotted logs and stumps, often in large numbers.

Hemitrichia clavata

Fruitbodies (sporangia) are clustered to crowded, ochre-yellow to mustard-yellow, 1–2 mm tall, clavate to pear-shaped, and with fluffy, cottony threads anchored in a deep cup (calyculus). Widespread and very common, *H. clavata* fruits on well-rotted wood. It is distinguished from *H. calyculata* (above) by its deep cup.

Hemitrichia serpula

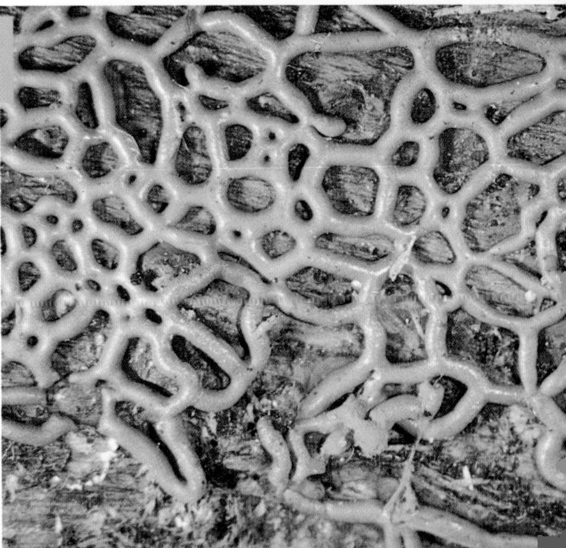

Fruitbody (plasmodiocarp) consists of a network of swollen, yellow to mustard-yellow veins that cover several to many square centimetres. Outer wall breaks down, and the internal material forms a cottony, mustard-yellow sporemass. Widespread but not common, this distinctive slime mould fruits on wood and plant debris.

Metatrichia vesparium

Fruitbodies (sporangia) are up to 3 mm tall, goblet-shaped, and wine-coloured to black-metallic, and fruit in clusters arising from a common stalk. Well-marked lid breaks down at maturity to expose a bright red-brown to purplish, fluffy mass of threads (capillitium, see below). Widespread and not uncommon, this species fruits on well-rotted wood.

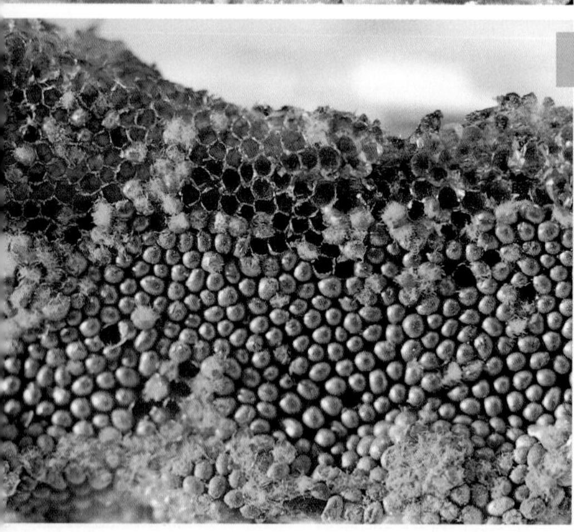

Trichia favoginea

Fruitbodies (sporangia) have no stalks (sessile) and are up to 2 mm tall by 0.5–1 mm wide, dull ochre to golden-brown, and produced in crowded masses, which might be several centimetres in extent. *En masse*, the sporangia can resemble a pseudoaethalium. Outer wall breaks down, and the mustard-yellow contents fluff up. Empty fruitbodies have a honeycombed appearance. Widespread and not uncommon, this slime mould fruits on well-rotted wood.

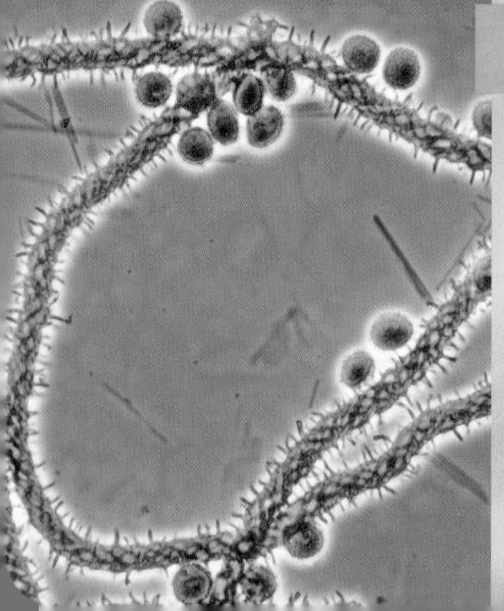

Slime Mould Capillitium

The fluffy material produced by species of the genera *Trichia* (above), *Hemitrichia* (p. 41), and *Metatrichia* (above) is called capillitium. It is composed of very fine threads, with spiral thickenings on the wall. The thickenings take up or lose water vapour, and the change in tension causes the threads to wriggle. The wriggling movement expels the tiny, spherical spores (seen as spheres in photo) into the air for dispersal. The threads are about 1/100th of a millimetre across (=10 µm).

Brefeldia maxima

Massive fruitbody (aethalium), a giant amongst the slime moulds, is up to 30 cm long. Young aethalium (inset photo) is white to off-white, with the texture of semolina pudding. At maturity, the aethalium turns sooty-black and is about 1.5 cm thick. Black, powdery spores are scattered when the outer crust breaks away. It is not immediately obvious, but this slime mould is related to *Stemonitis axifera* (below). Widespread but not common, this species fruits on dead wood and living plants.

Stemonitis axifera

Fruitbodies (sporangia) are up to 2 cm tall, cylindrical, and rusty-brown, and arise in dense clusters from a membranous base. Stalks are black, shining, and 3–7 mm tall. *S. axifera* fruits on dead wood. (There are several similar species that can only be distinguished by microscopic features.)

Comatricha typhoides

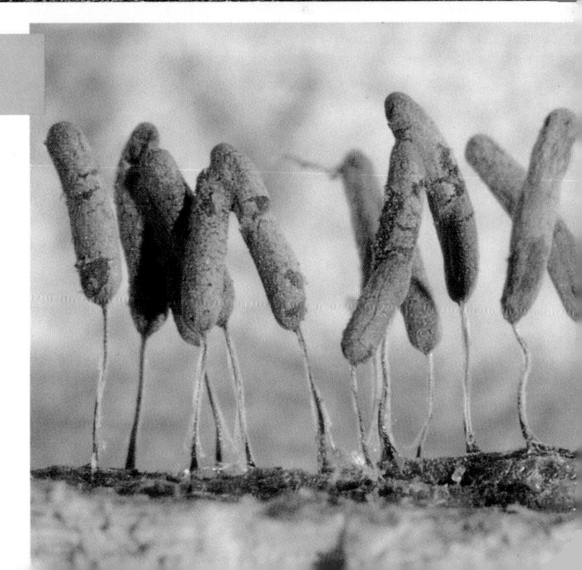

Fruitbodies (sporangia) are 2–5 mm tall, brown, and covered with a silvery outer wall when young. Wall breaks down to expose the brown spore-mass. *C. typhoides* looks like a short *Stemonitis axifera* (above), but the silvery wall is distinctive. Widespread and common, it fruits on logs, stumps, and woody debris.

Dictydium cancellatum

Chinese Lantern

Fruitbodies (sporangia) are 1–4 mm tall and with globose to flattened, red-brown to purple-brown heads. Outer wall has thick ribs, which persist to form a cage-like head at maturity. Chinese Lantern fruits on well-rotted wood. It is quite common, but difficult to spot because it blends in with the woody background.

Cribraria intricata

Fruitbodies (sporangia) are ochre to dusky-brown and up to 4 mm tall, with spherical heads about 0.5 mm wide. Head is composed of a network of fine threads, with prominent dark nodes at the junctions. Spore-mass is pale ochre. Widespread and common, *C. intricata* fruits on well-rotted wood.

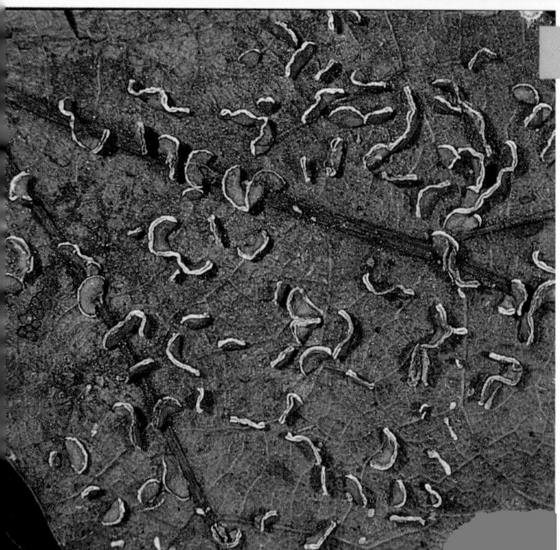

Physarum bivalve

Fruitbodies (sporangia) are variable in size from 1–10 mm long and 3–4 mm tall, very narrow, laterally compressed, clam-like, and vary from brown to dark grey. Open by a narrow slit that appears as a white, chalky line between the two valves. Sporangia fruit over the surface of fallen leaves and the like. This species is widespread but not common.

Lycogala epidendrum

Fruitbodies (aethalia) are 3–15 mm across, globose to subglobose or hemispherical, bright cinnabar-red when young, and ageing to sand-coloured or olivaceous-brown. Spore-mass is pinkish-grey to ochre. Common and widespread, *L. epidendrum* fruits scattered or crowded on well-rotted wood.

Lycogala flavofuscum

Fruitbodies (aethalia) are 2–4 cm across and silvery-grey to purplish-brown. Aethalia are much larger than those of *L. epidendrum*, and the wall is thick and brittle. Widespread but not common, this species fruits, scattered or in clusters, on well-rotted logs or stumps.

Enteridium lycoperdon

Fruitbodies (aethalia) are up to 8 cm across and cake-like, and have a silvery-brown outer wall. Base of the aethalium is surrounded by a conspicuous white margin (hypothallus). Spore-mass is dark brown. Widespread and common, this species migrates some distance before fruiting and is found in a wide variety of sites.

Tubifera ferruginosa

Fruitbodies (pseudoaethalia) are up to 15 cm long and cake-like. At maturity pseudoaethalia are red-brown to purplish-brown, with a metallic, iridescent surface. Spore-mass is umber-brown. Young sporangia are delicate pink to bright rose-red, and are amongst the most striking of the slime moulds, but they become drab and clay-brown at maturity. Widespread and common, this species fruits on well-rotted logs, often in moss.

Dictyaethalium plumbeum

Fruitbodies (pseudoaethalia) form a flat, cushion-like mass that is yellowish to olive-brown and up to 10 cm long by about 1 cm thick. At maturity, the caps of individual sporangia form a mosaic crust over the pseudoaethalium. Widespread and not uncommon, this species fruits on well-rotted wood.

SAC FUNGI

Ascomycota

This division is the largest of all the fungi and contains some of the most highly prized of the edible species (morels and truffles). Most sac fungi, however, are not important as food sources, because they are either poisonous or too rare, small, tough or flimsy.

While morels and truffles are treasured by gourmets, and other species of sac fungi are also considered edible and desirable, keep in mind that some sac fungi are **poisonous**. For many, edibility is still **unknown**. Thus, for the purposes of this guide, sac fungi are considered **not edible** unless otherwise noted. Interestingly, it is seldom reported that some of the morels are rather bland and not as "choice" as the gourmets would allow. So, if you are disappointed by your first morel, give them another try. False Morel, *Gyromitra esculenta* (p. 73), although classified as **poisonous**, is a popular edible fungus, especially in eastern Europe. While many people eat it with impunity, this fungus has been responsible for the **deaths** of hundreds of people over the years and should be avoided (see p. 319).

There are a number of sac fungi that also cause serious damage as parasites of garden ornamental plants (e.g., powdery mildews, see p. 29) or agricultural crops (e.g., ergot of rye). Many of the sac fungi are extremely interesting biologically, as wood rotters, as parasites (insects, plants or other fungi) or as saprobes on a wide range of organic materials. A few are known to be important in mycorrhizal associations with trees.

Sac fungi are so-called because the sexual spores (**ascospores**) are produced in a sac-like mother cell (**ascus**). The number of spores in each sac varies with different species, but is usually eight (see Fig. 3). Ascospores are extremely small and measured in micrometres, µm (1 mm=1000 µm). Spore-containing sacs are produced in large numbers (millions) in a fruitbody (**ascoma**), which can be one of three types: **cleistothecium**, **perithecium** and **apothecium**.

Fig. 3. **Ascus of Ascobolus.** The mature spores (right) are pigmented to protect them from ultraviolet radiation while they are airborne. The nucleus can be seen at the centre of each of the spores in the young ascus (left).

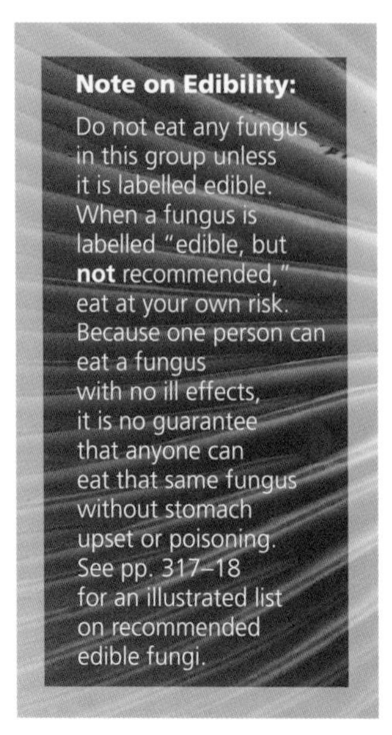

The **cleistothecium** is a tiny sphere, usually less than a millimetre in diameter, and is difficult to spot without a hand lens. Thousands of sacs with their spores are produced **inside** the cleistothecium, and these sacs are eventually released by the breakdown of the outside wall of the fruitbody. In your garden, you will often find the fruitbodies of powdery mildews, which appear as dark dots, on the undersurface of the leaves of numerous ornamental host plants. More than 30,000 species of flowering plants are attacked by powdery mildews!

The **perithecium** is a tiny, flask-shaped fruitbody, and the sacs with their spores are produced **inside**. Individual perithecia are usually less than 1 mm tall. Sometimes, however, they can be produced in such abundance that they are readily seen *en masse*, as in *Nectria* (p. 82). Often, perithecia are produced in large numbers embedded in a large fungal mass referred to as a **stroma**: in *Xylaria polymorpha* (p. 79), for example, the stroma is large (for a fungus) and gives rise to the common name of "Dead Man's Fingers." If you break *Xylaria* in two, you will see many tiny chambers lining the outside edge of the stroma. Each one of these chambers is a perithecium. Perithecia are usually embedded in the stroma with only the neck protruding. The protruding necks give the surface of the stroma a pimpled appearance. The spores are either shot out of the pore at the tip of the neck or ooze out like toothpaste.

The **apothecium** is the largest and most variable of the fruitbody types. Apothecia are shaped like plates, cups, saucers or urns. Some apothecia are clavate (club-shaped) or have stalked caps. The cup, saucer or disc (discoid) shapes are the classical apothecial forms, but the fruitbodies of morels, false morels, saddle fungi and the like are also considered apothecia. The key feature of sac fungi with apothecia is that the spore sacs are produced in a layer (**hymenium**) over the upper or outside facing surface of the fruitbody and are exposed to the air at maturity. A single fruitbody might produce hundreds of millions of sacs over its surface.

In sac fungi with apothecia, the spores are discharged violently: when ripe, a sac explodes and shoots its spores into the air above the fruitbody. Ascospores are so tiny that they are almost invisible. Sometimes, however, millions of sacs fire simultaneously into the air, and this cloud of hundreds of millions of spores appears as "smoke." This phenomenon is called "puffing." If you approach a cup fungus carefully and breathe on it gently, it might reward you by puffing. The tiny spores are carried great distances by wind to new sites for growth.

Fig. 4. DIAGRAM OF A CUP FUNGUS

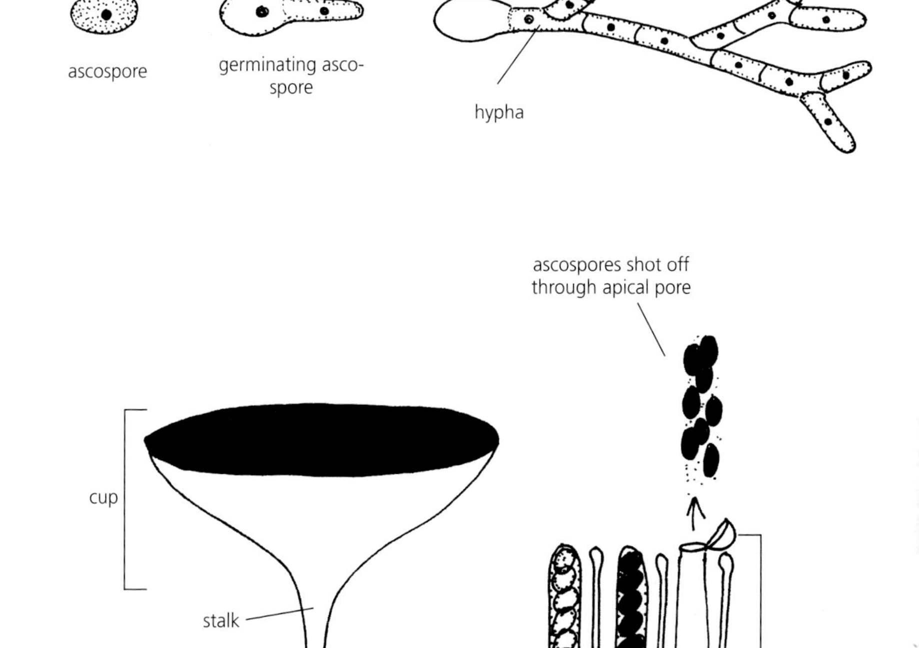

Fig. 4. **Ascomycota: Life Cycle of a Cup Fungus (Discomycete)**. The fertile layer (hymenium) lines the inside of the cup. The hymenium is composed of tightly packed asci separated by sterile cells. Each ascus contains eight spores, and these are shot off through a lid at the tip of the ascus. The spores are carried by the wind to new sites where they germinate to produce filaments (hyphae). The hyphae absorb nutrients, colonize the substrate and eventually produce more fruitbodies. The cup-shaped fruitbody is called an apothecium.

Key to Genera of Sac Fungi

FB = Fruitbodies

*genera not treated in this book

1. FB on wood or ground **2**
1. FB underground or on dung, mosses, or other fungi **73**

2. FB green, yellow-green, yellow, orange **3**
2. FB some other colour **19**

3. FB cup-shaped to discoid or ear-shaped. **4**
3. FB not as above **12**

4. Cups 1–4 cm or more in diameter **5**
4. Cups small (< 1 cm in diameter) **8**

5. Cups greenish to yellow/green .... ***Chlorencoelia***
5. Cups yellow to orange/yellow **6**

6. Cups more or less vertical, ear-shaped .... ***Otidea***
6. Cups horizontal **7**

7. Cups orange-yellow overall .... ***Aleuria***
7. Cups orange-yellow inside, green to olive outside .... ***Caloscypha***

8. Cups greenish to yellow-green .... ***Chlorencoelia***
8. Cups yellow to orange **9**

9. Cups orange, with a fringe of dark hairs .... ***Scutellinia***
9. Cups yellow **10**

10. Cups smooth .... ***Bisporella***
10. Cups hairy **11**

11. Cups covered with long, stiff, yellow hairs .... ***Dasyscyphus***
11. Cups covered with a layer of matted, white hairs .... ***Lachnellula***

12. FB tiny, 1–3 mm across **13**
12. FB much larger, clavate, irregular, or flattened **14**

13. FB tiny, orange spheres about 1 mm across .... ***Nectria***
13. FB irregular, pale lemon stromata up to 3 mm across .... ***Creopus***

14. FB yellow, variable in shape, flattened, stalk not obvious .... ***Neolecta***
14. FB with well-defined stalk and head **15**

15. Stalks and heads yellow to yellow-brown **16**
15. Stalks and heads different colours **17**

16. FB gelatinous, swelling when wet .... ***Leotia***
16. FB not gelatinous .... ***Microglossum, Cudonia***

17. FB gelatinous, stalks yellow, caps dark green .... ***Leotia***
17. FB not gelatinous, caps yellow, stalks white **18**

18. Heads elongate .... ***Mitrula***
18. Heads globose .... ***Vibrissia***

19. FB red to orange-red **20**
19. FB some other colour **21**

20. Cups stalked .... ***Sarcoscypha***
20. Stalks absent (sessile), with a fringe of dark hairs .... ***Scutellinia***

21. FB blue, blue/green, or green **22**
21. FB some other colour **24**

22. FB disc-, cup-, or fan-shaped **23**
22. FB clavate .... ***Microglossum, Leotia***

23. FB bluish .... ***Chlorociboria***
23. FB greenish .... ***Chlorencoelia***

24. FB violet to purple or violaceous-brown **25**
24. FB some other colour **29**

25. FB with white stalks .... ***Cudoniella***
25. FB purple or violaceous-brown, stalks absent **26**

26. FB thin, brittle, discoid, or cup-shaped .... ***Peziza***
26. FB strongly gelatinous **27**

27. FB deep purple .... ***Ascocoryne***
27. FB pinkish to brownish, with purple tinges **28**

28. FB pale pinkish to violaceous .... ***Neobulgaria***
28. FB irregular hemispherical .... ***Ascotremella***

29. FB white to off-white **30**
29. FB some other colour **36**

30. FB discoid or cup-shaped **31**
30. FB not as above **34**

31. Cups small (< 5 mm) **32**
31. Cups larger (1–10 cm or more) **33**

32. Cups smooth, on acorns or moss .... ***Hymenoscyphus***
32. Cups hairy, on wood .... ***Dasyscyphus***

33. Cups gelatinous, on wood .... ***Neobulgaria***
33. Cups thin, brittle, on ground .... ***Paxina costifera***

34. FB with fluted (lacunose) stalks, roughly saddle-shaped .... ***Helvella***
34. Stalks not obvious, shape columnar or irregular **35**

35. FB columnar, like stalk of mushroom .... ***Underwoodia***
35. FB appear as tiny dots in malformed mushroom .... see ***Hypomyces***

36. FB grey to black **37**
36. FB tan, buff, brown to red-brown, or blackish-brown **51**

37. FB tiny (1 mm), globose, clustered .... ***Lasiosphaeria***
37. FB much larger **38**

38. FB cup-shaped or discoid **39**
38. FB not as above **42**

39. FB thick-walled, rubbery to gelatinous .... ***Bulgaria***
39. FB not as above **40**

40. FB deeply goblet-shaped .... ***Urnula***
40. FB flat or bowl-shaped **41**

41. FB black, tough .... Plectania*
41. FB grey to brown or dull black, fragile .... ***Helvella***

42. FB black, hard, tough **43**
42. FB black, stalked, rubbery to brittle **48**

43. FB thread-like, finger-like, or flattened, white inside .... ***Xylaria***
43. FB hemispherical or crust-like, black inside **44**

44. FB hemispherical **45**
44. FB flattened, black crust **46**

45. FB 1–5 cm, several, concentric zones inside .... ***Daldinia***
45. FB > 1 cm, lacking zones .... ***Hypoxylon***

46. Crust extensive, irregularly bumpy .... Ustulina*
46. Crust more or less smooth **47**

47. Crust thin .... ***Hypoxylon***
47. Crust thick .... Camarops*

48. FB clavate, with a flattened head **49**
48. Not as above **50**

49. FB smooth over entire surface .... ***Geoglossum***
49. FB roughened (hand lens) with hairs .... ***Trichoglossum***

50. FB elongate with a pimply surface, rubbery .... ***Cordyceps***
50. FB with fluted (lacunose) stalks, brittle .... ***Helvella***

51. FB disc-, cup-, or ear-shaped **52**
51. FB not as above **63**

52. FB red-brown, on scales of fir cones .... Ciboria*
52. FB not as above **53**

53. FB on wood **54**
53. FB on ground **56**

54. FB thick-walled, brown, rubbery .... ***Galiella***
54. FB not as above **55**

55. FB smooth, discoid, coppery-brown .... ***Pachyella***
55. FB cup-shaped, coated with dark brown hairs .... ***Humaria***

56. Outside covered with dark brown hairs .... ***Geopora***
56. Outside of FB smooth **57**

57. FB lacking stalks **58**
57. FB stalked **61**

58. Cups more or less vertical or ear-shaped .... ***Otidea, Helvella***
58. Cups horizontal **59**

59. Upper surface of disc wrinkled .... ***Discina,*** Disciotis*
59. Upper surface of disc smooth **60**

60. FB shallow cups, average >1.5 cm .... ***Peziza***
60. FB deep cups, average < 1.5 cm .... ***Tarzetta***

61. FB arising from a buried sclerotium .... Sclerotinia*
61. FB not associated with sclerotium **62**

62. Stalks longer than width of disc .... ***Helvella***
62. Stalks short and stubby .... ***Tarzetta***

63. FB above ground **64**
63. FB wholly or partially underground **73**

64. FB tan-coloured crust on wood or old conk .... ***Hypocrea***
64. FB not as above **65**

65. FB flattened .... ***Spathulariopsis, Spathularia***
65. FB not flattened **66**

66. FB with honeycombed cap .... ***Morchella***
66. Caps not honeycombed **67**

67. FB with saddle-shaped caps .... ***Gyromitra, Helvella***
67. Caps not saddle-shaped **68**

68. Caps wrinkled, convoluted, or fluted **69**
68. Caps more or less clavate **71**

69. FB convoluted, brain-like .... ***Gyromitra***
69. FB with vertical wrinkles or flutes **70**

70. FB with more or less vertical flutes .... ***Mitrula***
70. FB with vertical wrinkles .... ***Ptychoverpa***

71. Heads buff, 3 mm tall, supported by wiry stalks .... ***Heyderia***
71. FB with elongate caps **72**

72. FB with pimpled heads .... ***Cordyceps***
72. FB with smooth heads .... ***Microglossum***

73. FB underground, blackish-brown, often associated with Cordyceps .... ***Elaphomyces***
73. FB on dung, mosses, catkins, or other fungi **74**

74. Attacking mushrooms that turn bright yellow or orange .... ***Hypomyces***
74. FB on herbivore dung, mosses, or catkins **75**

75. FB on herbivore dung .... ***Peziza***
75. FB on mosses, parasitic .... ***Hymenoscyphus***

Aleuria aurantia

Orange Peel

Fruitbodies (apothecia) are up to 8 cm across, bright orange, cup-shaped at first, and becoming flattened with a wavy margin in age. Flesh is thin and brittle. Widespread and common, this distinctive cup fungus fruits on bare or disturbed soil at the edges of trails or gravel roads in summer and fall. It lacks the stem of *A. rhenana* (below). Edible.

Aleuria rhenana

False Orange Peel

Fruitbodies (apothecia) are up to 3 cm across, cup-shaped, and bright orange. Cups have well-developed stalks that distinguish False Orange Peel from *A. aurantia*. It grows in tight clusters, often amongst mosses, in coniferous woods. Although widespread, this species is more common in western North America.

Chlorociboria aeruginascens

Blue-Stain Fungus

Fruitbodies (apothecia) are up to 1 cm across, cup-shaped to fan-shaped, bright to dark blue-green, and attached by a short stalk. Widespread and common, Blue-Stain Fungus fruits on old, rotting logs and twigs. The fruitbodies only appear during prolonged wet weather, but the fungus is still visible at other times because the blue hyphae permeate the wood. A very similar species (*C. aeruginosa*) is distinguished microscopically.

Chlorencoelia versiformis

Fruitbodies (apothecia) are up to 1.5 cm across, cup-shaped to ear-shaped or fan-shaped, greenish to yellow-green or pale olive-green, and with a short stalk. Widespread but not common, this species fruits on hardwood logs in summer.

Sarcoscypha austriaca

Scarlet Cup

Fruitbodies (apothecia) are cup-shaped to discoid, up to 4 cm across, brilliant red inside, whitish outside, and with a short, whitish stalk. Common and widespread, it fruits on twigs or small branches. One of the earliest of spring fungi, it often escapes attention because it is hidden under fallen leaves. The similar *S. coccinea* is found in the Pacific Northwest, but not in eastern North America.

Sarcoscypha occidentalis

Stalked Scarlet Cup

Fruitbodies (apothecia) are vermilion-red, saucer-shaped to trumpet-shaped, and with a disc up to 1 cm across. Stalks are white to reddish and up to 3 cm tall. Widespread and common, this striking cup fungus fruits on sticks or small branches in spring and early summer.

Caloscypha fulgens

Fruitbodies (apothecia) are up to 5 cm long by 4 cm across, cup-shaped to boat-shaped, and bright orange-yellow inside when fresh and staining greenish in age. Greenish to olive-green tints on the outside fade with age. Widespread and common, this attractive cup fungus fruits early in spring under conifers.

Ascocoryne sarcoides

Fruitbodies (apothecia) are up to 1 cm across, gelatinous, and violet to purple or reddish-purple. Young fruitbodies appear as purple lobes bursting out of the wood. Lobes absorb water and expand to form cups or discs, which swell and coalesce to form a gelatinous mass. Widespread and common, this purple sac fungus fruits on stumps or logs of hardwoods.

Ascocoryne cylichnium

This species is very similar to *A. sarcoides*, but the fruitbodies tend to be more discoid and less gelatinous. The two species can only be separated with certainty by the spores, which are 10–19 µm by 3–5 µm in *A. sarcoides* and 18–30 µm by 4–6 µm in *A. cylichnium*. Widespread, this species fruits on hardwood stumps and logs.

Ascotremella faginea

Fruitbodies (apothecia) are swollen, gelatinous, purplish to smoky-brown, with violet tints, tightly clustered, and form a complex, lobed mass, up to 4 cm across and several centimetres tall. As the genus name suggests, this sac fungus looks like a jelly fungus (see *Tremella*, p. 104). Widespread but not common, this species fruits on debarked hardwoods, particularly beech and alder.

Bulgaria inquinans

Fruitbodies (apothecia) are up to 4 cm across, black, rubbery when fresh, leathery when dry, smooth to top-shaped, and with a flat or slightly concave disc. This sac fungus is widespread, but not common and fruits on fallen trunks and branches of hardwoods.

Photo: Brian Shelton.

Galiella rufa

Hairy Rubber Cup

Fruitbodies (apothecia) are up to 3 cm across, shallow cup-shaped, tan to brownish, with thick, rubbery flesh, and anchored by a short stalk. Outer wall is hairy, wrinkled, and dark brown. Hairy Rubber Cup is widespread but not common and fruits on hardwood twigs or branches in spring and summer.

Neobulgaria pura

Fruitbodies (apothecia) are 1–2 cm across, top-shaped, soft, gelatinous, violaceous to flesh-coloured or pinkish, and with a flat or slightly concave disc. Widespread and not uncommon, this species fruits in dense clusters on hardwood branches or logs.

Dasyscyphus virgineus

Fruitbodies (apothecia) are cup-shaped to discoid, up to 1.5 mm across, and with cream-coloured discs. Outside of the cup is covered with a dense mat of white hairs. In dry weather the cup margin rolls in, and the hairs fold over to protect the disk and reduce water loss. In wet weather, the margin unrolls to expose the disc. Widespread and common, this species fruits over the surface of dead twigs, raspberry canes, etc.

Dasyscyphus sulphureus

Fruitbodies (apothecia) are up to 2 mm across and cup-shaped to discoid. Outer surface is covered with stiff, bright yellow hairs. In dry periods, the hairy outer surface folds over to protect the disc and in wet weather, it opens up to expose the hymenium. This sac fungus is widespread, but is often overlooked because of its small size. It fruits on the dead stems of herbaceous plants.

Lachnellula agassizii

Fruitbodies (apothecia) are discoid, bright yellow to orange-yellow, and up to 5 mm across. Outer wall is fringed with a dense mat of white hairs. This species fruits in large numbers on the bark of dead conifers. It is not uncommon, but is best seen during prolonged wet periods when its bright colour attracts the eye. Also known as *Dasyscyphus agassizii*.

Photo: Greg Thorn.

Bisporella citrina

Lemon Drops

Fruitbodies (apothecia) are bright yellow, saucer-shaped to discoid, smooth, and up to 3 mm across. Stalks are short and inconspicuous. This species is one of the commonest of the woodland cup fungi and fruits in large numbers, sometimes hundreds, on decorticated logs or branches of hardwoods.

Scutellinia scutellata

Eyelash Fungus

Fruitbodies (apothecia) are discoid, up to 1 cm across, and bright red to orange-red or orange, with a fringe of stiff, black hairs around the margin. Widespread and common, this cup fungus fruits on well-rotted wood, scattered or in clusters.

Scutellinia setosa

Fruitbodies (apothecia) are saucer-shaped to discoid, 2–3 mm across, crowded, and dull orange to orange-yellow, with a fringe of dark hairs around the margin. Widespread and not uncommon, it fruits on well-rotted deciduous wood. Also known as *S. erinaceus*.

Humaria hemisphaerica

Fruitbodies (apothecia) are cup-shaped, up to 3 cm across, and whitish inside. Outer wall is covered with stiff, dark brown hairs, which also form a fringe around the edge of the cup. This sac fungus is widespread, but not particularly common and fruits on the ground or on well-rotted wood.

Peziza praetervisa

Fruitbodies (apothecia) are cup-shaped, thin, brittle, and become flattened with a wavy outline in age. Inside wall is violet to violet-brown, but changes to brown with age, and the outside wall is paler. Widespread and not uncom mon, it fruits, scattered or in clusters, amongst charred wood in campsite fireplaces.

Peziza domiciliana

Fruitbodies (apothecia) are cup-shaped to discoid, thin, brittle, up to 10 cm across, pale flesh-coloured to beige, ageing brown, and with a whitish exterior. Inner and outer surfaces are both smooth. Widespread and common, it fruits on wet ground, dirt floors of garages or basements, and sometimes (as in this case) through cracks in concrete floors.

Peziza repanda

Spreading Cup

Fruitbodies (apothecia) are buff to tan, cup-shaped to saucer-shaped at first, becoming flattened and irregularly wavy in age, and up to 12 cm or more across. Widespread and common, Spreading Cup fruits on rotting logs or woody debris.

Peziza ammophila

Sand-Loving Cup

Fruitbodies (apothecia) are up to 4 cm wide, dark brown inside, with an ochre outer surface, and cup-shaped at first, but split and reflex to become flattened on the sand in a star-shaped form. Cups are partially immersed and anchored by a tuft of fungal threads. This species fruits in the coastal areas and around the Great Lakes region in fall.

Photo: Greg Thorn.

Peziza vesiculosa

Fruitbodies (apothecia) are 3–7 cm wide, deeply cup-shaped or long boat-shaped, pale brown to greyish brown, and often with an inrolled margin and a mealy exterior. Widespread and common, this species fruits alone or often in clusters on the ground in woods or open places or sometimes on animal droppings.

Tarzetta cupularis

Fruitbodies (apothecia) are up to 1.5 cm across, brittle, pale tan-coloured, and goblet-shaped, with a finely toothed margin. Outer wall is covered with tiny, brown pustules. Short stalks (up to 4 mm) are often buried in the soil. Widespread but not common, this species fruits partially buried in the ground in woods.

Discina perlata

Fruitbodies (apothecia) are up to 8 cm across, brown, discoid, brittle, and with an irregularly bumpy upper surface. Widespread and not uncommon, this species fruits in spring or early summer on bare ground under conifers or near conifer stumps at the edge of woodland paths. (*Disciotis venosa* fruits about the same time in the same locations, but is larger and has raised veins over the upper surface of the disc.) *Discina perlata* is also known as *Gyromitra perlata*.

Geopora sepulta

Fruitbodies (apothecia) are up to 7 cm across, deeply cup-shaped, smoky-brown to red-brown outside, with dark hairs, and dirty white to yellowish inside. They are nearly buried in soil and often split at the edges and reflex to give an irregular star-shaped outline. Widespread and common, this cup fungus fruits partially buried in the ground in woods. Also known as *Sepultaria sumneriana*.

Urnula craterium

Devil's Urn

Fruitbodies (apothecia) are black, up to 12 cm tall, 3–8 cm across the circular mouth, goblet-shaped, with a toothed margin, and supported by a slender stalk. Stalks are often immersed in moss. Fruitbodies often escape attention because they are buried in the duff. Widespread and not uncommon, Devil's Urn arises from buried wood to fruit on the ground in spring.

Pachyella clypeata

Copper Penny

Fruitbodies (apothecia) are discoid and lack a stalk. Discs can be flat or wrinkled and fluted, and are 1–2.5 cm across. Upper surface is dull or shiny and red-brown to copper-coloured. Widely distributed and not uncommon, Copper Penny is found closely attached to well-rotted logs in summer and fall.

Otidea auricula

Ear-Shaped Otidea

Fruitbodies (apothecia) are up to 6 cm tall by 5 cm wide, ear-shaped, and with vertically orientated cups that are split lengthwise with overlapping edges. Cups are rich red-brown inside and somewhat paler outside, with a whitish base. Widespread but not common, this sac fungus fruits under conifers. Based on microscopic features, this species is now referred to as *Helvella silvicola* by some mycologists.

Otidea onotica

Orange Otidea

Fruitbodies (apothecia) are up to 10 cm tall, thin, fragile, irregularly ear-shaped, and orange to ochre-yellow, with a pinkish tinge. Widespread and not uncommon, Orange Otidea fruits on the ground in clusters in deciduous woods in fall.

Otidea leporina

Hare's Ear

Fruitbodies (apothecia) are up to 10 cm tall by 5 cm wide, yellow-brown, vertically orientated, with inrolled edges, and shaped like the ear of a rabbit. Widespread but rare, Hare's Ear fruits in groups or clusters on the ground.

Hymenoscyphus fructigenus

Acorn Cup Fungus

Fruitbodies (apothecia) are tiny, discoid, up to 4 mm across, and white to off-white or cream-coloured. Stalks are whitish and variable in length. This sac fungus is widespread and fruits on beechnuts and acorns during prolonged wet periods.

Hymenoscyphus subcarneus

Fruitbodies (apothecia) are tiny (less than a mm across) and consist of a thin, translucent, whitish disc. This species attacks moss gametophytes that turn brown and die and on which the fungus then produces the tiny, scattered discs. The healthy, dark green moss can be seen in close juxtaposition to the brown, infected moss (in photo).

Leotia lubrica

Common Jelly Baby

Fruitbodies (apothecia) are gelatinous, yellow to yellow-brown, and up to 5 cm tall, with an irregularly lobed head, 1–2 cm across. Stalks are cylindrical and dull yellow-brown. Widespread and common, this jelly baby is found in groups or dense clusters on the ground in the woods during wet periods.

Leotia viscosa

Green-Capped Jelly Baby

L. viscosa is similar to *L. lubrica* in its size, general appearance, and gelatinous fruitbody. However, this jelly baby is easily recognized by its bright yellow stalk and green to blackish-green head. Widespread and common, it fruits in wet spots or on old, moss-covered stumps.

Microglossum rufum

Yellow Earth Tongue

Fruitbodies (apothecia) are clavate, flattened, up to 3 cm tall, with a cylindrical, roughened stalk, and bright yellow to orange-yellow. Widespread and common, Yellow Earth Tongue fruits on the ground in wet spots or on very rotten wood. At first glance, it looks like a coral fungus (see *Clavaria*, pp. 111–112), but is related to the earth tongues and produces asci, not basidia.

Microglossum olivaceum

Olive Earth Tongue

Fruitbodies (apothecia) are clavate, up to 6 cm tall, pale to dark olive-brown, smooth, and flattened, with a central groove. Widespread but rare, Olive Earth Tongue fruits on the ground in mixed woods.

Mitrula elegans

Fruitbodies (apothecia) are up to 4 cm tall, with elongate, bright yellow to pinkish-yellow, slimy heads and whitish, cylindrical stalks. In early spring, this striking species is found fruiting in large numbers in standing pools of water. *Vibrissea truncorum* (p. 70) is similar, but prefers to fruit in running water and is usually submerged.

Photo: Greg Thorn.

Geoglossum difforme

Common Earth Tongue

Fruitbodies (apothecia) are clavate, up to 10 cm tall, flattened, black, and sticky to slimy when wet. Widespread and common, Earth Tongue fruits on the ground or on well-rotted wood. There are a number of similar species that can only be distinguished from each other by microscopic features.

Trichoglossum hirsutum

Hairy Earth Tongue

Fruitbodies (apothecia) are black, clavate, and ellipsoid to elongate. They are similar to the fruitbodies of Common Earth Tongue (top photo), but in Hairy Earth Tongue the head and stalk are covered with stiff, black hairs that give it a roughened outline (readily seen with a hand lens). Hairy Earth Tongue is the most common of a number of hairy species that can be separated only by microscopic characteristics.

Vibrissea truncorum

Fruitbodies (apothecia) are up to 12 mm tall. Caps are yellow to orange-yellow and hemispherical. Stalks are downy to irregularly bumpy and white at first, but become dotted with black tufts, especially near the base. Widespread and not uncommon, this fungus fruits in spring and early summer on twigs and branches in streams.

Cudonia circinans

Fruitbodies (apothecia) are 1.5–6 cm tall and pale brown to ochre, with thin flesh. Caps are up to 2 cm across and overhang the cylindrical stalks. This fungus resembles Common Jelly Baby (*Leotia lubrica*, p. 67), but has thin flesh and is not gelatinous. Widespread and not uncommon, *C. circinans* fruits on forest debris or on very rotten wood.

Spathularia flavida

Yellow Fairy Fan

Fruitbodies (apothecia) are up to 4 cm tall by 2 cm wide (at the widest point). Head is fan-shaped to spathulate and ochre to yellow-brown. Widespread but not common in eastern North America, this species fruits under conifers.

Spathulariopsis velutipes

Velvet-Stalked Fairy Fan

Fruitbodies (apothecia) are up to 6 cm tall. Heads are flattened, fan-shaped to spathulate, and pale yellow-brown to dark brown. Stalks are cylindrical, brown, darker than the head, and minutely hairy to velvety. Widespread and locally common, Velvet-Stalked Fairy Fan fruits on the ground or on well-rotted hardwood. It differs from Yellow Fairy Fan in having a darker, somewhat velvety stalk. Also known as *Spathularia velutipes*.

Neolecta irregularis

Fruitbodies (apothecia) are up to 7 cm tall by 3 cm wide, bright yellow, clavate to spathulate or irregularly lobed, and occasionally branched. Surface is bumpy or wrinkled. At first glance, this sac fungus looks like a coral fungus (see *Clavaria*, pp. 111–112). Widespread and not uncommon, this species fruits in wet, mossy spots in woods. Reported to be edible, but **not** recommended.

Morchella elata

Black Morel

Fruitbodies (apothecia) are up to 20 cm tall. Heads are dark brown to brownish-black, narrowly conical, and up to 10 cm tall by 8 cm wide at the base. Heads have deep chambers in rows separated by dark ridges. Stalks are off-white to cream, roughened (mealy), and up to 10 cm tall. Widespread and common, Black Morel fruits on the ground in woods in spring. Edible.

Morchella esculenta

Yellow Morel

Fruitbodies (apothecia) are up to 15 cm tall. Heads are subglobose to ellipsoid or broadly conical, pale brown to yellow-brown, honeycombed, and with well-defined, irregularly arranged pits. Stalks are white to buff, delicately roughened, and broadest near the base. Widespread and common, Yellow Morel fruits on the ground in woods in spring. Edible.

Morchella semilibera

Half-Free Morel

Fruitbodies (apothecia) are up to 15 cm tall. Heads are brown to dark brown, up to 5 cm tall, bell-shaped to broadly conical, and with elongate, fluted chambers. Margin of the cap is skirt-like and free from the stalk. Stalks are off-white to yellowish-white, roughened, and up to 10 cm tall. Widespread and common, Half-Free Morel fruits under hardwoods in spring. Also known as *Mitrophora semilibera*. Edible.

Gyromitra esculenta

False Morel

Fruitbodies (apothecia) are up to 23 cm tall. Heads are dark brown, up to 8 cm across, folded and wrinkled, brain-like, and red-brown to blackish-brown. Stalks are paler than the cap and up to 15 cm tall. Widespread and locally common, False Morel fruits on the ground in deciduous and coniferous woods in spring. It is often eaten, but has caused numerous deaths (see p. 319). **Poisonous.**

Photo: Greg Thorn.

Ptychoverpa bohemica

Wrinkled Thimble Cap

Fruitbodies (apothecia) are yellow-brown to dark brown. Caps are broadly conical to bell-shaped, up to 5 cm tall by 5 cm wide, and with tight vertical wrinkles. Cap hangs skirt-like, with the bottom margin free from the stalk. Stalks are cylindrical and off-white to ochre. Widespread and common, this morel fruits on the ground in woods in spring. Also known as *Verpa bohemica*. Edible, but it sometimes causes **stomach upset**.

Helvella crispa

White Elfin Saddle

Fruitbodies (apothecia) are white to off-white, up to 10 cm tall, and with irregularly saddle-shaped heads. Stalks are fluted (lacunose). Widespread but not common, White Elfin Saddle fruits solitary or scattered, either on the ground or on very rotten logs in coniferous woods in summer and fall.

Helvella lacunosa

Black Elfin Saddle

Fruitbodies (apothecia) are up to 15 cm tall. Caps are grey-brown to dull black, with scalloped lobes. Stalks are white to smoky and deeply fluted (lacunose). Widespread and common, Black Elfin Saddle fruits on the ground in woods. Edible.

Helvella elastica

Common Elfin Saddle

Fruitbodies (apothecia) are up to 10 cm tall, with thin, brittle flesh. Caps are dark grey to smoky-brown, lobed or saddle-shaped, and with an inrolled margin. Stalks are smooth, slender, cylindrical, and white. Common Elfin Saddle is widespread and fruits on the ground in woods.

Gyromitra infula

Saddle-Shaped False Morel

Fruitbodies (apothecia) are up to 10 cm across, brown to red-brown, more or less saddle-shaped, and covered with shallow, irregular folds. Stalks are up to 10 cm long and white to brown. Widespread and common, Saddle-Shaped False Morel fruits in fall on soil rich in woody debris or on well-rotted logs.

Helvella costifera

Fruitbodies (apothecia) are up to 6 cm tall by 6 cm wide, cup-shaped at first, but becoming irregularly flattened in age, and white to off-white. Stalks are smooth, white, and fluted (lacunose). This relatively rare fungus fruits under conifers. Also known as *Paxina costifera.*

Helvella macropus

Fruitbodies (apothecia) are cup-shaped to discoid, yellow-brown to grey-brown, and 2–4 cm across. There are no veins on the outside wall and the cup is supported by a slender, cylindrical stalk. Superficially unlike most of its "saddle fungus" relatives, this species fruits on the ground or on very rotten wood and is widespread and common.

Helvella queletii

Fruitbodies (apothecia) are cup-shaped to saucer-shaped, with an inrolled margin, 3–6 cm wide, and about as tall. Inside is smooth and grey-brown to blackish-brown. Exterior is whitish to ochre-grey. Stalks are up to 6 cm long, cylindrical, and slightly swollen towards the base. Widespread but infrequent, this species fruits on the ground, often on hard-packed soil at the edges of trails.

Hypoxylon fragiforme

Beech Hypoxylon

Hard, brittle, hemispherical mass (stroma), up to 1.5 cm across, forms on dead beech wood. Stromata are dull brick-red to black, with a pimpled surface (seen with a hand lens). Pimples are the openings of tiny fruitbodies (perithecia) embedded in the stroma. Beech Hypoxylon is widespread and common, especially on logs and branches of beech.

Hypoxylon multiforme

Birch Hypoxylon

Hemispherical or irregularly shaped crust (stroma) forms over the surface of the bark of birch logs and branches. Crusts are red-brown when young, but become black in age, hard, tough, and can persist throughout the year. Tiny, embedded fruitbodies (perithecia) of the fungus appear as pimples over the surface of the crust.

Daldinia concentrica

King Alfred's Cakes

Hard, brittle, red-brown to shiny black, hemispherical mass (stroma), up to 4 cm across, forms on stumps and decorticated logs. When split, the stroma shows clearly defined concentric zones. Tiny, flask-shaped cavities (perithecia) open to the exterior and give a pimpled appearance to the outer surface of the fruitbody. *Hypoxylon multiforme* (p. 78) is similar, but lacks internal zonation. King Alfred's Cakes is widespread and common.

Xylaria polymorpha

Dead Man's Fingers

Stroma is clavate to finger-like or irregular, 2–8 cm tall by 2 cm or more wide, and cylindrical or flattened. Outside is hard and black and inside is white. Surface is pimpled, revealing the mouths of the tiny, embedded fruitbodies (perithecia). Widespread and common, Dead Man's Fingers fruits in clusters at the base of, or near, rotting stumps. Note: *Xylaria* is pronounced "Zylaria!"

Xylaria longipes

Stalked Xylaria

Fungus forms clavate structures (stromata), up to 5 cm tall by 1 cm wide. Stromata are greyish-white to brown on the outside, becoming black in age, and white inside. Outside wall is often mottled. Stalk is slender, cylindrical, and of variable length. Widespread and common, Stalked Xylaria fruits on maple or sometimes beech logs or branches.

Xylaria hypoxylon

Candlesnuff

Fruitbodies are 2–7 cm tall, flattened, and tough, with antler-like branching. Branches are black, with white tips. White powder at the tips is the asexual spore state for summer dispersal. Sexual state of this fungus (top right) is found later in the year and forms black, slender, pimply, often unbranched stalks. Sexual state produces thick-walled spores that help the fungus survive difficult times such as winter and drought. Widespread and common, Candlesnuff fruits on or around dead wood.

Lasiosphaeria spermoides

Fruitbodies (perithecia) are black, about 0.5 mm across, more or less globose, and with short, pointed beaks. Although large for perithecia, the fruitbodies are still difficult to spot because of their small size and dark colour. Widespread but seldom reported, this fungus fruits in dense masses on wood or bark.

Cordyceps militaris

Stromata are bright orange to pale orange, with flattened, clavate heads. Heads are 2–4 cm tall and the surface is distinctly pimpled. These are the mouths of tiny, embedded fruitbodies (perithecia). Widespread but infrequent, this species is parasitic on insect pupae and, with careful digging, the mummified body of the host (bottom of photo) can be found.

Cordyceps ophioglossoides

Adder's Tongue

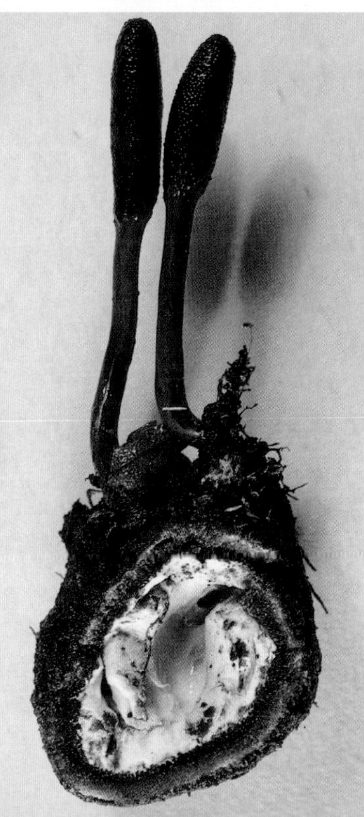

Fungus forms an elongate, clavate structure (stroma), 2–10 cm tall, and with a dark brown, pimpled head. Stalk is usually paler than the head. Widespread and not uncommon, Adder's Tongue is parasitic on the fruitbodies of an underground fungus called *Elaphomyces* (below).

Elaphomyces granulatus

Fruitbodies are globose to subglobose, up to 4 cm across, pale ochre, but becomes blackish, and warty. Interior is purplish and becomes black as the spores mature. Fruitbodies are found under the ground and fruit on conifer woods or hardwoods. The fungus can be exposed by raking the soil, but a patch is often discovered by finding *Cordyceps* spp. (see above and p. 82).

Cordyceps capitata

Round-Headed Cordyceps

Fruitbodies are stalked, up to 10 cm tall, with spherical, pimpled heads that are brown to blackish-brown, and up to 1 cm across. Stalk is yellowish. Fungus is widespread but not common and parasitizes an underground fungus called *Elaphomyces* (p. 81). Fruitbodies are found alone or in small groups under conifers in fall. If you dig below the Round-Headed Cordyceps, you will find the *Elaphomyces* host.

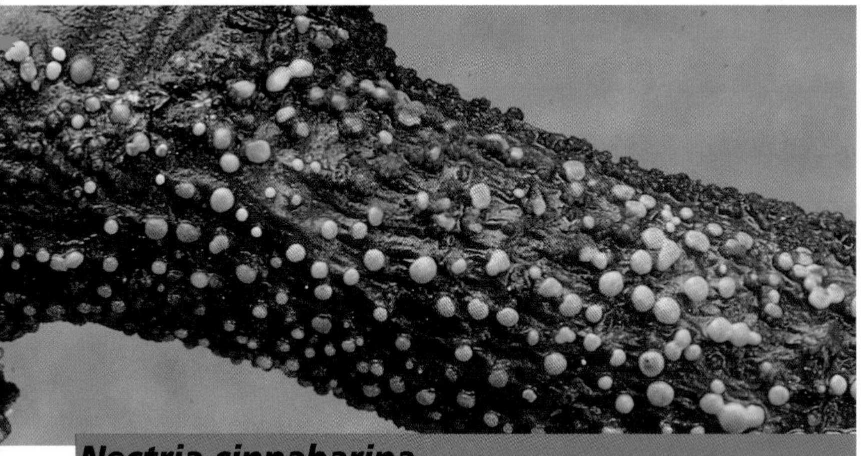

Nectria cinnabarina

Coral Spot

Coral Spot attacks a number of hardwood hosts as a weak parasite. It also grows prolifically on dead twigs and branches of hardwoods. The flesh-coloured spots are the summer spore state and the orange spots are clusters of tiny fruitbodies (perithecia) that produce the hardy overwintering spores. This fungus is a common backyard one.

Underwoodia columnaris

Fruitbodies are up to 15 cm tall, columnar, with longitudinal flutes, and dirty white to tan, but becoming brown in age. When cut, large internal chambers are revealed. *U. columnaris* is not obviously a sac fungus and might be mistaken for a parasitized mushroom. Spores are produced in a layer over the outside surface. Widespread but rare, it fruits on the ground in woods in early summer.

Photo: Greg Thorn.

Hypomyces lactifluorum

Lobster Mushroom

This parasite attacks the fruitbodies of milk mushrooms (*Lactarius* spp.). Diseased host mushrooms become brilliant orange or orange-red, are malformed, and the gills either do not develop or are reduced to low ridges. Inspection with a hand lens reveals tiny, dark fruitbodies (perithecia) of the parasite embedded in the host mushroom. Widespread and common, Lobster Mushroom is **reported** to be edible.

Hypomyces chrysospermus

Whitish at first, diseased host mushrooms turn bright golden-yellow as they are overrun by this parasite. *H. chrysospermus* produces massive numbers of spores that appear like yellow dust over the surface of the host. It is a common and widespread parasite, especially of boletes, during prolonged wet periods.

Hypocrea pulvinata

Pale yellow-brown to ochre crust (stroma) forms on dead bracket fungi, this species' substrate. Tiny fruitbodies (perithecia) are embedded in the stroma. Mouths of the perithecia appear as dark brown dots scattered over the surface of the crust. Widespread but not common.

Hypocrea gelatinosa

Stromata are small, discrete, and up to 3 mm across, containing embedded fruitbodies (perithecia). Stromata are at first pale lemon-yellow and translucent. Spores are green and as they mature the mouths of the perithecia appear as dark green dots scattered over the stroma. Widespread but not common, this species fruits in numbers on well-rotted wood. Also known as *Creopus gelatinosus*.

Heyderia abietis

Caps are pale yellow-brown, cylindrical to clavate, up to 3 mm tall by 2 mm wide, and are supported by a smooth, slender, dark brown stalk up to 3 cm tall. This sac fungus fruits on conifer needles. It is common but rarely reported, and is perhaps overlooked because of its small size.

BASIDIOMYCOTA

In this division, the spores (usually four) are borne externally on a clavate (club-shaped) spore mother cell called a **basidium**. In the Class Hymenomycetes, the basidia are packed into a layer called a **hymenium**. The purpose of the hymenium is to give a large surface area for spore production—the greater the surface area, the greater the number of spores produced—and the gills, tubes, spines, etc., found on fruitbodies are designed to maximize the hymenial surface.

(Fig. 5 shows a section through a wedge-shaped gill typical of most of the gill fungi, pp. 179–315.) At maturity of the fruitbody, the spores are shot off a very small distance (0.1–0.2 mm), sideways from the basidium, into the space between the gills. The spores then drop by gravity below the bottom edge of the gill, where they are picked up by air movements for dispersal. The Division Basidiomycota is broken up into a number of smaller groups called classes. Classes end with the suffix *-mycetes*. In this book, the two most important classes in the Basidiomycota are the Hymenomycetes and the Gasteromycetes. The Hymenomycetes have a hymenium as described above and the spores are shot off. In the Gasteromycetes the spores mature **inside** the fruitbody and are released in a number of unusual ways as outlined below.

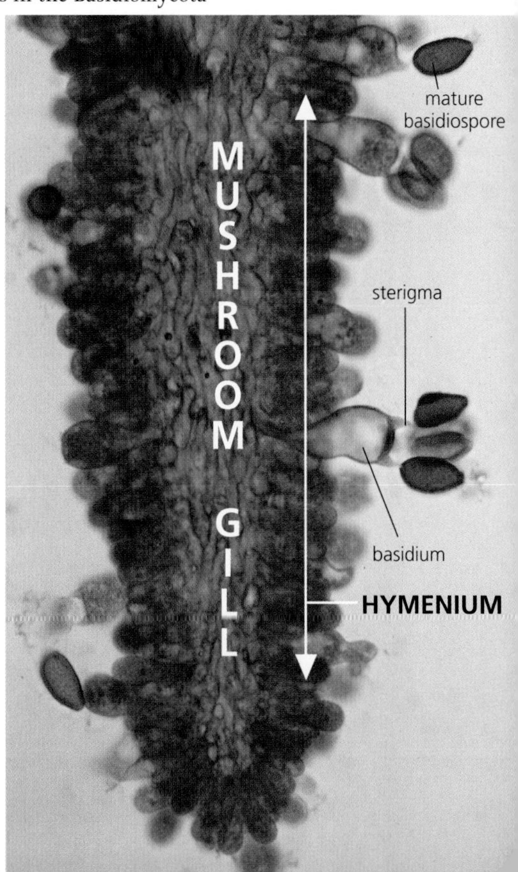

Fig. 5. **Section of a Mushroom Gill.** The spore mother cell in Basidiomycota is called a basidium. Usually four basidiospores are produced on each basidium. Each basidiospore is produced at the apex of a pointed prolongation called a sterigma. In mushrooms, the basidia are packed into a layer that lines the face of the gill called a hymenium. In the photomicrograph (at right), there are some mature basidia in the hymenium with the spores on the prongs ready to be discharged. When it is ripe, each spore is shot off a very small distance (0.2 mm) between adjacent gills. It then falls by gravity below the bottom level of the gills and is carried off by air currents for dispersal.

PUFFBALLS AND FRIENDS

Basidiomycota—Gasteromycetes

Puffballs, earthstars, bird's nest fungi, stinkhorns and related fungi produce their spores **inside** the fruitbodies, instead of discharging the spores directly into the air. For this reason, these fungi are placed in the class Gasteromycetes (stomach fungi). Members of this group are very variable in size and shape and have many interesting and novel ways of spreading their spores using wind, insects or rain. Gasteromycetes range from Giant Puffball (*Calvatia gigantea*, p. 93), weighing many kilograms, to tiny *Sphaerobolus stellatus* (p. 98), which is only 2 mm across.

Puffballs and Earthstars

Puffballs are spherical, subspherical, ellipsoid, pestle-shaped or pear-shaped. The outer wall is usually roughened with spines that flake off to reveal a smooth, membranous inner wall. The spores mature inside the fruitbody as a powdery mass. In Giant Puffball, the outer wall cracks and breaks off as it dries out, exposing the spore-mass to the elements for dispersal. In the smaller puffballs (*Lycoperdon* spp., pp. 88–90), there is a well-defined pore at the apex. When raindrops strike the outside wall of the spore sac, the implosion of the wall forces a puff of spores out through the pore, for dispersal by the wind.

In the earthstars (*Geastrum* spp., pp. 95–96), the thick outer wall splits and the segments form a number of pointed arms. As the fruitbody dries, the arms reflex to expose a puffball-like spore sac. Earthstars form on the forest floor, just below the loose duff. The reflexing arms raise the inner spore sac above the ground for better dispersal. The spores are puffed out through a pore by raindrops in the same way as the smaller puffballs.

Bird's Nest Fungi

The fruitbody of a Bird's Nest Fungus (BNF) looks like a tiny nest with eggs. The "eggs" (**peridioles**) are packages of thousands of spores contained within a hard outer wall (see Fig. 6). In some BNF the eggs are anchored to the side wall by a structure that contains a long, thread-like tail (**funiculus**), with a sticky base (**hapteron**). Falling raindrops cause mini-explosions in the cone-shaped cups, and the splash propels the eggs out of the cup. Eggs can be shot nearly 2 m away from the cup, and they attach to a suitable substrate by means of the sticky base.

The "Big Bertha" of the fungal world is *Sphaerobolus stellatus* (p. 98). About 2 mm across, *S. stellatus* catapults its solitary, spherical egg (1 mm in diameter) a distance of more than 6 m!

Note on Edibility:

Do not eat **any** fungus in this group unless it is labelled edible. For further information, see pp. 317–18 for an illustrated list on recommended edible fungi.

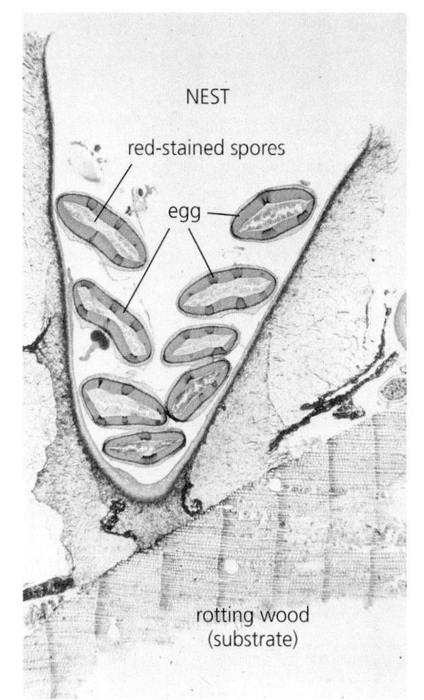

Fig. 6. **Fruitbody of a Bird's Nest Fungus** (BNF).

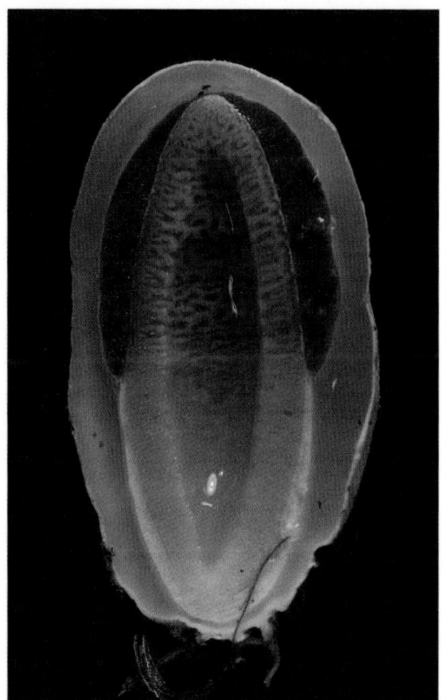

Fig. 7. **Stinkhorn Egg**.

Fig. 6. **Fruitbody of Bird's Nest Fungus** (BNF). Section through the fruitbody of *Crucibulum* shows the "eggs" inside the "nest." The outside layers of the egg are stained greenish. The "egg" in a Bird's Nest Fungus is a packet containing thousands of spores. These are stained red and are massed at the centre of each egg. Eggs are splashed out of the cup by raindrops and are propelled a distance of nearly 2 m.

Fig. 7. **Stinkhorn Egg**. Section through the egg of *Mutinus* shows the various layers. The gelatinous layer immediately below the outside wall mixes with the green spore-mass beneath it as the stalk expands. The mixture forms the evil-smelling goo attractive to flies. In the bottom half of the egg you can see the compact column (columella) that expands to become the porous stalk. When conditions are right, the egg can expand to full size within 30 minutes.

Stinkhorns

In their young stage, the "egg stage," stinkhorns resemble puffballs, and their fruitbodies are spherical to subspherical or ellipsoid. Their internal organization, however, is more complex (see Fig. 7): a gelatinous layer surrounds an olive-green spore-mass that covers the head of the stinkhorn. In the eggs of *Phallus* spp. and *Mutinus* spp., the head surrounds a central column. At maturity, the wall cracks open and the column expands to form the support stalk. At this time, the gelatinous layer mixes with the spore-mass to produce a fetid, evil-smelling, olive-green goo that also contains sugary materials. The strong odour attracts flies from great distances to feed on the sweet, smelly stuff, which sticks to the flies' body parts. Because the spores are mixed in with the goo, the flies unwittingly transport the spores to other likely sites.

Lycoperdon curtisii

Curtis's Puffball

Fruitbodies are globose to some what flattened and up to 2 cm across. They are white at first, ageing brown, and have persistent spines. Spore-mass is olive-brown. Common and widespread, Curtis's Puffball fruits in grassy places in small clusters.

Lycoperdon pusillum

Fruitbodies are 1–2 cm across, white and powdery at first, becoming smooth and clay-coloured, and changing to brown in age. They lack a sterile base and are globose to subglobose. Widespread and not uncommon, this species produces a few scattered fruitbodies on bare ground in pastures or open woods.

Lycoperdon candidum

White Puffball

Fruitbodies are up to 4.5 cm tall. Head is up to 3 cm across and has a short, rooting base. Outer wall is covered with pyramidal spines that slough off in thick patches to reveal a brown to olive inner wall. Spore-mass is olive-brown. Widespread and common, White Puffball fruits on sandy soil. It is also known as *L. marginatum*. Edible.

Lycoperdon nigrescens

Black Puffball

Fruitbodies are up to 8 cm tall by 2–5 cm wide at the top, pear-shaped, umber, ageing to dull brown or tan, and with a well-developed, sterile base. Outer layer of spines is composed of closely packed hairs. Spines slough off to leave a network of scars on the wall (areolate). Widespread and locally common, Black Puffball fruits on the ground in woods. Also known as *L. foetidum*. Edible.

Lycoperdon perlatum

Gem-Studded Puffball

Fruitbodies are up to 6 cm tall by 5 cm wide, pear-shaped to top-shaped, with a sterile, stalk-like base, and white at first, ageing yellow-brown to brown. Outer wall forms cone-shaped spines that slough off to leave a network of scars (areolate). Spore-mass is olive-brown. Widespread and common, Gem-Studded Puffball usually fruits in clusters on the ground in humus or on woody debris. Edible.

Lycoperdon pyriforme

Pear-Shaped Puffball

Fruitbodies are up to 5 cm tall and 3 cm wide, pear-shaped, with a sterile base, white to tan or red-brown, and minutely roughened at first, but becoming smooth and papery in age. Pore is well-marked, but is often slow to form. Spore-mass is olive-brown. This species is our most common puffball, and it fruits in dense clusters on rotting logs or stumps. Edible, but lacks flavour.

Photo (bottom): Gerald Stephenson.

Morganella subincarnata

Fruitbodies are up to 3.5 cm across, usually wider than tall, and with spines grouped to form dark, pyramidal warts that disappear in age to leave a pitted or netted surface. Widespread but not common, this species fruits scattered or in dense clusters on rotten logs. Also known as *Lycoperdon subincarnata*. It differs from Pear-Shaped Puffball (p. 90) because it lacks a sterile base.

Tulostoma campestre

Stalked Puffball

Spore sac is 1–1.5 cm across, buff-coloured, with an apical pore surrounded by a darker zone, and is perched on top of a long, cylindrical stalk. Stalks are up to 5 cm tall and anchored by yellowish fungal strands. Stalk is buried and is not seen, unless dug out or exposed by weathering. Spores are puffed through the pore as in *Lycoperdon* spp. (pp. 88–90). Widespread but not common, Stalked Puffball fruits in sandy soil.

Photo: Greg Thorn.

Astraeus hygrometricus

Water Measurer

Fruitbodies are up to 5 cm across, with 6–15 acutely pointed arms. Spore sacs have an irregularly shaped pore. Arms are ornamented with a network of fissures (areolate), and are open when wet but close over the spore sac when dry. Widespread and locally abundant, Water Measurer fruits in sandy areas, especially around the Great Lakes and in coastal regions.

Scleroderma citrinum

Earthball

Fruitbodies are 2.5–10 cm across, globose to subglobose or ellipsoid, yellow-brown to golden-brown, and with a thick outer wall, usually ornamented with a distinctive raised mosaic. At maturity, fruitbodies rupture irregularly to form a crater-like opening and expose the purple-brown spore-mass. Common and widespread, Earthball fruits on or around old stumps in wet spots. It is sometimes attacked by *Boletus parasiticus* (p. 162). **Poisonous.**

Scleroderma areolatum

Fruitbodies are 1–6 cm across, irregularly globose to ellipsoid, and anchored by a short base or thick tuft. Wall is thin, fragile (compared with Earthball, above), and ornamented with flattened, dark brown scales. Surface is areolate (covered with a network of fissures). Spore-mass is purple, ageing to olive-brown. Spores are spiny and 10–18 μm in diameter. Widespread, it is often found on humus or under shrubs in ornamental gardens. **Poisonous.**

Calvatia gigantea

Giant Puffball

Fruitbodies are globose to subglobose, ellipsoid or irregular, and very large (up to 50 cm or more across). Outer wall is smooth and white when young, ageing to tan or cinnamon and breaking away to expose the spore-mass. Widespread and common, Giant Puffball fruits in the rich soil of fields, woods, gardens or on the banks of streams. Also known as *Langermannia gigantea*. Edible.

Photo: John Sutton.

Calvatia cyathiformis

Fruitbodies are up to 15 cm tall by 10 cm wide (at their widest part), pear-shaped, and with a globose to subglobose head. Fruitbodies are white at first, ageing brown, smooth, becoming cracked over the surface, and wrinkled where the head joins the stalk. Stalks are thick and tapering. Spore-mass is rich purple-brown. Widespread but not common, this species fruits in grassy spots. Edible when **young**.

Bovista plumbea

Tumbling Puffball

Fruitbodies are globose to subglobose and 2–4 cm across, and open by a large, circular pore. White, fuzzy outer wall wears off to reveal a grey to purple-brown, metallic inner wall. At maturity, the puffball breaks free and is blown around in fields and other grassy places. The similar *B. pila* is bronze at maturity. Tumbling Puffball is widespread and not uncommon.

Calvatia excipuliformis

Pestle-Shaped Puffball

Fruitbodies are up to 15 cm tall by 10 cm across and pestle-shaped, and open by an irregular rupture. They are white at first, ageing brown, and are often wrinkled where the head meets the stout stalk. Spore-mass is dark brown. Widespread and not uncommon, Pestle-Shaped Puffball fruits on the ground in woods and could be mistaken for a small puffball (*Lycoperdon* spp., pp. 88–90), but it is larger and lacks the well-defined pore characteristic of that genus. Edible when **young**.

Geastrum quadrifidum

Four-Armed Earthstar

Fruitbodies are up to 4 cm across. Spore sacs are purple, taller than wide, with a short support stalk, and raised above the ground by 4–5 reflexed arms. Spore sac has a clearly defined, finely striate mouth area (peristome). Widespread but not common, Four-Armed Earthstar fruits on the ground under conifers.

Geastrum pectinatum

Beaked Earthstar

Fruitbodies have 5–10 arms, which reflex to raise the spore sac into the air. Spore sac is globose to subglobose, about 2.5 cm across, and with an elongate, deeply grooved, beak-like mouth (peristome) that is clearly demarcated from the rest of the spore sac. Beaked Earthstar fruits on the ground or on well-rotted stumps, and is not common.

Geastrum fimbriatum

Fringed Earthstar

Fruitbodies are up to 6 cm across, with 5–9 reflexed arms. Spore sacs are 1–2.5 cm across and nest in the reflexed arms. Mouth (peristome) is delicately fringed at the apex and, unlike Nested Earthstar (p. 96), is not clearly defined from the remainder of the spore sac. Widespread but not common, Fringed Earthstar fruits under conifers or hardwoods.

Geastrum triplex

Collared Earthstar

Fruitbodies are large, 5–10 cm across, and tan to reddish-brown. Spore sacs nest in the basal cup formed by the reflexed arms. Arms are fleshy, often splitting as they reflex to form a collar around the base of the spore sac. Mouth area (peristome) is clearly defined and striate. Widespread and fairly common, Collared Earthstar, which is large for an earthstar, fruits on the ground under hardwood trees.

Geastrum saccatum

Nested Earthstar

Fruitbodies are small, 1–4 cm across, buff to tan, and with 5–8 reflexed arms. Spore sacs are 0.5–2 cm across and nest in the reflexed arms. The striate mouth (peristome) is clearly defined from the remainder of the spore sac. Nested Earthstar does not possess a collar as in Collared Earthstar (above). Widespread and not uncommon, small Nested Earthstar fruits on the ground under conifers.

Crucibulum laeve

White Bird's Nest Fungus

Fruitbodies are up to 1 cm tall by 1 cm wide, broadly cone-shaped, velvety, and tan-yellow to cinnamon. Inside of the nest is smooth, shining, and whitish. Eggs (peridioles) are whitish to cream-coloured and about 1.5 mm across. Common and widespread, White Bird's Nest Fungus fruits on dead twigs, leaf mould, rotting wood, and the like.

Cyathus stercoreus

Dung Loving Bird's Nest

Fruitbodies are up to 1.5 cm tall by 4–8 mm wide, cone-shaped, and buff to golden-brown or almost black in age. They are woolly on the outside and lead-grey to blue-black, but not striate, on the inside. Widespread and fairly common, Dung Loving Bird's Nest fruits on herbivore dung, manured soil, wood chips, and the like.

Cyathus olla

Grey Bird's Nest

Fruitbodies are up to 1.5 cm tall by 1 cm wide, brownish, and hairy. With age, they become smooth and grey, with a flaring mouth and a wavy margin. Inside is smooth and metallic-grey to dull black. Eggs (peridioles) are large and up to 3.5 mm across. Widespread and common, Grey Bird's Nest fruits on organic debris, wood chips, etc. It is often found fruiting on old husks in cornfields.

Cyathus striatus

Striate Bird's Nest

Fruitbodies are up to 1 cm tall by 4–6 mm wide, cone-shaped, and covered with shaggy, dark brown hairs. Inside wall is smooth, shining, striate, and lead-grey. Widespread and not uncommon, Striate Bird's Nest fruits on woody debris.

Sphaerobolus stellatus

Fruitbodies are tiny, 1–2 mm across, and whitish to straw-coloured or orange-yellow. Outer wall splits and reflexs to form a tiny star, exposing the spherical "egg" (peridiole). Egg is catapulted a distance of more than 6 m. Widespread and not uncommon, *S. stellatus* fruits on decomposing plant material.

Dictyophora duplicata

Skirted Stinkhorn

Fruitbodies are up to 20 cm tall, with a chambered head (morel-like), and covered with an olive-green spore-mass. Net-like skirt flares out below the head. Stalks are stout, porous, and white. Egg stage is 4.5–7 cm across, white, subglobose, and with shallow depressions over the surface. Widespread and not uncommon, Skirted Stinkhorn fruits in rich soil in woods and gardens.

Photo: Brian Shelton.

Phallus ravenelii

Ravenel's Stinkhorn

Fruitbodies are up to 10–15 cm tall. Head is covered by a slimy, dark olive-green spore-mass and is smooth to granular underneath. Spore-mass is never chambered, distinguishing Ravenel's Stinkhorn from *P. impudicus*, which has a chambered (morel-like) head after the spore-mass is removed. Widespread and not uncommon, this species fruits on well-rotted logs or woody debris, sawdust, etc.

Photo: Brian Shelton.

Mutinus ravenelii

Dog Stinkhorn

Fruitbodies are up to 10 cm tall. Head is up to 2.5 cm long, conical, with a reddish tip, and covered with an olive-green spore-mass. Stalks are pink to reddish, porous, and about 1 cm wide. Common and widespread, Dog Stinkhorn fruits in rich soil in gardens or woods. *M. caninus* is similar, but has a slender, whitish stalk.

Lysurus cruciatus

Lizard's Claw

Fruitbodies are up to 10 cm tall. Stalks are white, porous, and bear a claw-like cluster of 4–6 hollow arms at the apex. Spore-mass is dark olive-green. This distinctive species fruits on the ground in plant debris, sawdust, and occasionally cornfields. Lizard's Claw is widespread but not common. Also known as *Anthurus borealis*.

JELLY FUNGI

Basidiomycota—Phragmobasidiomycetes

As the common name suggests, the fruitbodies of most jelly fungi are in a gelatinous matrix. In dry weather, fruitbodies lose water to form irregular, horny masses that shrivel and almost disappear. In rainy weather, the gelatin absorbs water rapidly and the fruitbodies recover their normal shape, size and colour and resume spore production. This ability to revive allows jelly fungi to persist for many weeks and explains why they appear so quickly after rain.

The jelly fungi group is not as species rich as other groups of fungi. A few, such as Orange Jelly (*Dacrymyces palmatus*, p. 102), are both colourful and common. For the most part, jelly fungi are too small, too scattered or too infrequent to be important as edibles. There are exceptions, however, and Apricot Jelly Fungus (*Tremiscus helvelloides*, p. 107) is appreciated by some. Although it does not have much flavour, the delicate pink/apricot/orange colouring makes it an attractive dressing in salads. You can also find cans or bags of white jelly fungus (*Tremella* spp.) imported from the Orient on the grocer's shelf or in Chinese or Japanese specialty food stores. Again, they are an interesting food additive but, for flavour, do not have much to offer.

Some fungi belonging to the "jelly group" (Tremellales) are not obviously gelatinous. For example, False Coral Fungus (*Tremellodendron pallidum*, p. 108) has dry, tough, flattened branches. Microscopically, it is related to jelly fungi, but it looks more like a coral fungus. Unlike most coral fungi, however, its branches are tough, persistent and almost woody in texture. Yellow Staghorn Fungus (*Calocera viscosa*, p. 103) also resembles a coral fungus, but this species is also tough and persistent, and it can revive in wet weather. In coral fungi, on the other hand, the fruitbodies are fragile, break easily and decay quickly. The delicate, fragile nature of a coral fungus is the easiest way to distinguish it from a jelly fungus. For the specialist, microscopic characteristics show that in *Tremellodendron* spp. the spore mother cell is partially divided into four compartments, while in *Calocera* the spore mother cell is tuning fork–shaped. In true coral fungi, the spore mother cell is undivided and clavate (see Fig. 5, p. 85). In mycology, these differences are regarded as important.

Note on Edibility:
Do not eat **any** fungus in this group unless it is labelled edible. For further information, see pp. 317–18 for an illustrated list on recommended edible fungi.

The true jelly fungi belong to the Basidiomycota (the spores are borne externally on basidia). Some sac fungi (Ascomycota), such as *Ascotremella* (p. 56) and *Neobulgaria* (p. 57), mimic the true jellies and also produce fruitbodies in a gelatinous matrix, which is an example of how unrelated organisms can evolve to the same endpoint. In appearance, *Ascotremella* is unlike most other sac fungi and resembles a typical jelly fungus. Microscopic analysis, however, confirms that the spore mother cells are asci (the spores are formed internally), not basidia (the spores are formed externally).

Dacrymyces palmatus

Orange Jelly

Fruitbodies are 2–6 cm long by 1–3 cm tall, slimy, soft, gelatinous, orange to orange-yellow, spathulate, but becoming lobed or multi-lobed, and convoluted. Common and widespread, Orange Jelly fruits on dead conifer logs and stumps. This species can be confused with Witch's Butter (below), but it commonly fruits on conifers and is more common in eastern regions than Witch's Butter. It is distinguished microscopically from Witch's Butter by the tuning-fork basidia.

Tremella mesenterica

Witch's Butter

Fruitbodies are 2–8 cm long, irregular, lobed, gelatinous, yellow, fading to pale yellow or whitish, drying to a horny mass, and reviving when wetted. Widespread but not as common in the east as usually indicated, Witch's Butter fruits on twigs, logs, etc., of hardwoods during prolonged wet periods.

Calocera cornea

Fruitbodies are 2–12 mm tall, slender, unbranched or sparingly branched, tough gelatinous, dull yellow, and often with browning tips. This species dries to a horny, brown mass but revives when wetted. Common, it fruits on decorticated hardwood in scattered clusters or often in rows through cracks.

Calocera viscosa

Yellow Staghorn Fungus

Fruitbodies are 3–8 cm tall, yellow to orange-yellow, coral-like but tough and gelatinous, with open branching, and short, pointed laterals. Widespread and not uncommon, Yellow Staghorn Fungus fruits on dead conifer logs and stumps or sometimes on well-rotted wood after rising from the ground. This species is distinguished from a coral fungus by its tough, gelatinous texture.

Dacryopinax spathularia

Fan-Shaped Jelly Fungus

Fruitbodies are 5–20 mm tall, bright orange to orange-yellow, smooth, gelatinous, and consisting of fan-shaped to spathulate tongues that emerge from decorticated wood. Widespread but not common, Fan-Shaped Jelly Fungus usually fruits in clusters or in dense rows through cracks in the wood.

Tremella foliacea

Leafy Jelly Fungus

Fruitbodies are variable in size, from 5–25 cm long, tough, and gelatinous, and form densely packed, flattened lobes that are pale red-brown to deep red-brown, often with purple tinges. Widespread but not common, Leafy Jelly Fungus fruits on dead branches of hardwood trees and occasionally conifers.

Tremella reticulata

White Coral Jelly Fungus

Fruitbodies are up to 8 cm tall and 5–12 cm wide, consisting of a tightly packed cluster of tubular branches that form a hemispherical or elongate, rosette-like structure. Branches are white to off-white or pale cream, tubular, and radiating. Widespread and not uncommon, White Coral Jelly Fungus fruits on the ground in hardwood forests.

Tremella concrescens

Fruitbodies are white, gelatinous, and variable in size, consisting of a white, gelatinous growth. They grow from the ground and envelop and cling to living plant parts, such as the stems and leaves of adjacent weeds and grasses. This species is widespread and not uncommon during and after prolonged wet periods.

Auricularia auricula

Ear Fungus

Fruitbodies are 2–10 cm across, irregularly cup-shaped to ear-shaped, rubbery to gelatinous, red-brown to violet-brown, and often with raised, vein-like markings on the inside of the "ear." Outside is slightly hairy. Inside is smooth. Fruitbody becomes horny and black when dry but revives when rewetted. Widespread and not uncommon, Ear Fungus fruits on dead branches and logs. Edible.

Heterotextus alpina

Fruitbodies are cup-shaped to cone-shaped, with narrow attachment points, 5–15 mm across and about the same in height, yellow to orange-yellow, tough, and gelatinous. Widespread and locally common, it fruits on dead twigs, branches, and stumps of conifers. Also known as *Guepiniopsis alpina*.

Exidia glandulosa

Black Witch's Butter

Fruitbodies are olive-brown to black and form a series of cone-shaped, gelatinous bodies that expand, with wetting, to form an irregular mass of coalescing, flattened discs in narrow rows up to 25 cm long. Widespread and not uncommon, Black Witch's Butter fruits on twigs and branches of hardwoods.

Photo: Greg Thorn.

Exidia alba

White Jelly Fungus

Fruitbodies are 1–10 cm across, gelatinous, lobed, convoluted, often fusing in brain-like masses to cover the length of a log, white, becoming ivory, sometimes tinged with pink or violet, and staining purplish to brownish in age. Widespread but not common, White Jelly Fungus fruits on debarked hardwood.

Pseudohydnum gelatinosum

Toothed Jelly Fungus

Caps are 2–5 cm across, white to grey or grey-brown, rough to velvety on top, spongy-gelatinous, and shelving to spathulate. Undersurface of the cap is white and lined with densely packed, translucent teeth. Stalks are sometimes lacking or up to 5 cm long and lateral. Widespread and common, Toothed Jelly Fungus fruits on well-rotted conifer wood.

Tremiscus helvelloides

Apricot Jelly Fungus

Fruitbodies are 2–10 cm tall and up to 6 cm wide, tongue-shaped to spathulate or funnel-shaped, smooth, pink to apricot or reddish-orange, and gelatinous to rubbery. Stalks are not well differentiated from the cap. Widespread and fairly common, Apricot Jelly Fungus fruits on the ground under conifers or sometimes on well-rotted wood. Also known as *Phlogiotis helvelloides*. Edible, but without flavour.

Tremellodendropsis semivestitum

Fruitbodies are up to 6 cm tall, consisting of a stalk up to 2 cm tall by 3 mm wide, branching tree-like at the apex, and whitish to straw-coloured or pale tan. Branches are flattened and tough, and tips are often blackish with age. Widespread and common although seldom reported, this species fruits on the ground in woods. Also known as *Lachnocladium semivestitum*.

Tremellodendron pallidum

False Coral Fungus

Fruitbodies consist of a cluster of tightly packed branches up to 15 cm wide by 10 cm tall. Branches are tough, flattened, dry, not gelatinous, and off-white to cream. Widespread and common in some regions, False Coral Fungus fruits on the ground under hardwoods. Also known as *T. schweinitzii*. Its tough consistency distinguishes it from the true coral fungi (pp. 110–20).

Syzygospora mycetophila

Parasitic Jelly

Fruitbodies are 2 mm to 2 cm or more across and equally high, gelatinous, and pale tan. They form convoluted, brain-like masses on infected fungi. Widespread and common, Parasitic Jelly fruits on the caps and stalks of *Gymnopus dryophila* (p. 256) during wet periods. Several related species can only be distinguished microscopically. Also known as *Christiansenia mycetophila*. Reported as edible but tasteless.

Physalacria inflata

Bladder Fungus

Fruitbodies are tiny, 5–20 mm tall, white to pale cream, and with an inflated head that is supported by a slender stalk. Head is more or less globose, up to 10 mm across, smooth, white to cream, hollow, easily dented, and becoming flattened. Widespread and not uncommon during wet periods, Bladder Fungus fruits on well-rotted wood. This species is not one of the jelly fungi, but could be mistaken for a member of this group.

CORAL FUNGI

Basidiomycota—Hymenomycetes

In **Coral Fungi**, the spores are produced by a layer (hymenium) of spore mother cells (basidia) that covers the outside surface of the fruitbody. Fruitbodies grow **upwards**, either as simple stalks or branching, coral-like growths. This upwards growth of fertile branches distinguishes coral fungi from tooth fungi (pp. 121–28), where the spore-producing layer covers the outside of **downwardly** projecting spines.

In *Clavaria* spp. (pp. 111–112), the fruitbody is unbranched, more or less erect and worm-like or clavate (club-shaped). These simpler forms are sometimes referred to as **Club Coral**, and they resemble some of the sac fungi that are club-shaped (see pp. 68–69). More often, however, the fruitbodies of coral fungi form coral-like masses. Sometimes the branching is open, as in *Ramariopsis kunzei* (p. 119), and other times it is compact, as in Straight-Branched Coral (*Ramaria stricta*, p. 119). Coral fungi can be drab brown, but they are often lightly or brightly coloured and range from white to tan or bright orange-yellow to rose or purple.

Note on Edibility:

Do not eat **any** fungus in this group unless it is labelled edible. For further information, see pp. 317–18 for an illustrated list on recommended edible fungi.

Coral fungi are an attractive group, but are often difficult to identify to species, and sometimes even to genus, because the criteria used to distinguish them are based on microscopic features. For example, *Clavaria* spp. are distinguished from *Clavulina* spp. by the number of spores on the spore mother cells (basidia).

Many of the coral fungi are good edibles; others, such as Pink-Tipped Coral (*Ramaria formosa*, p. 117), produce gastrointestinal **toxins**. A few "edible" species can cause **upsets** in sensitive individuals. Coral fungi are **poor** candidates to eat for these reasons, as well as for the difficulties in their identification.

As noted above, some of the simple club-shaped coral fungi are very similar in physical appearance to the club-shaped sac fungi. Coral fungi tend to be soft, fragile, often brittle and decay quickly, whereas sac fungi are tougher and more resilient and persist for much longer before decaying. Similarly, some of the jelly fungi (e.g., Yellow Staghorn Fungus, p. 103) can be easily mistaken for coral fungi, but, as with sac fungi, jelly fungi are tough and can persist for weeks or even months, to revive again in wet weather.

Clavaria purpurea

Purple Club Coral

Fruitbodies are unbranched, very brittle, up to 6.5 cm tall by 2–6 mm wide, and purple to smoky-purple or pale violet. Widespread and not uncommon, Purple Club Coral is found in scattered clusters and sometimes dense stands in wet, grassy spots. Edible.

Clavaria vermicularis

Worm-Like Coral

Fruitbodies are unbranched, up to 6 cm tall, chalk-white, brittle, cylindrical, slender, with a short, translucent stalk, and erect or sometimes bent and worm-like. Widespread and common, Worm-Like Coral fruits scattered or in clusters in grassy spots in open woods. Edible.

Clavaria rosea

Rosy Club Coral

Fruitbodies are unbranched, fragile, up to 6 cm tall by 5 mm wide, and rose-pink. Stalks are translucent, white to pinkish, and not well-marked. Widespread but not common, Rosy Club Coral fruits on the ground, often beside woodland paths.

Clavaria fumosa

Fruitbodies are unbranched, up to 12 cm tall by 6 mm wide, fragile, in dense clusters, and pale flesh-coloured to pale grey, with a yellowish tinge. Widespread and locally common, this species fruits on the ground in hardwood forests.

Macrotyphula juncea

Fairy Thread

Fruitbodies are unbranched, 2–8 cm tall, very slender, yellowish to ochre, and produced in scattered stands. They are erect at first, but often become bent and twisted. Widespread and locally common, Fairy Thread fruits on leaves in damp spots in deciduous woods. Also known as *Clavariadelphus juncea*.

Clavaria ornatipes

Fruitbodies are unbranched to sparingly branched, up to 4 cm tall by 1.5 cm wide, greyish to pinkish-grey, tough, and with a hairy stalk. This species will shrivel when dry and revive when moistened. Widespread but not common, it fruits on the ground in woodlands.

Clavariadelphus ligula

Strap-Shaped Coral

Fruitbodies are unbranched, up to 7 cm tall by 3–12 mm wide, salmon-buff to brownish, and with a smooth or wrinkled surface. Clubs are flattened, cylindrical or spindle-shaped, often wider near the tip, and narrowly spathulate. Spore print is white to yellowish. Widespread and common, Strap-Shaped Coral fruits on the ground in coniferous or mixed woods.

Clavariadelphus truncatus

Flat-Topped Coral

Fruitbodies are thick and fleshy, up to 15 cm tall, yellow to yellow-brown, and clavate, with a distinctive, flattened (truncate) apex and frequently with longitudinal wrinkles. Widespread but not common, Flat-Topped Coral fruits under conifers. Edible.

Clavariadelphus pistillaris

Pestle-Shaped Coral

Fruitbodies are 6–15 cm tall by 1–4 cm wide, fleshy, clavate, red-brown, and roughened or wrinkled near the apex. Widespread but not common, Pestle-Shaped Coral fruits on the ground in mixed woods. Edible.

Clavulina cristata

Cockscomb Coral

Fruitbodies are branched, up to 8 cm tall and wide, chalk-white, ageing dirty white or greyish, and often blackish at the base (because of a parasitic fungus). Branches are characteristically flattened (truncate) near the apex and toothed, like a cockscomb (cristate). Very common and widespread, Cockscomb Coral fruits on the ground in woody debris. Edible.

Clavulina cinerea

Grey Coral

Fruitbodies are branched, up to 10 cm tall and wide, pale grey to blue-grey, and with paler, acutely pointed tips. Grey Coral is considered by many to be a variety or diseased specimen of Cockscomb Coral (above). Common and widespread, it fruits on the ground in woods. Edible.

Clavulina rugosa

Fruitbodies are 4–12 cm tall, white to off-white, and with a well-defined stalk and sparse, irregular, knobby branches. Considered by many to be a variety of Cockscomb Coral (above). Widespread and not uncommon, it fruits on the ground in woods, often beside footpaths. Edible.

Clavulinopsis fusiformis

Spindle-Shaped Coral

Fruitbodies are unbranched, up to 10 cm tall, bright yellow, hollow, cylindrical or flattened, and often narrower at the base and apex (spindle-shaped). Fruitbodies have a bitter taste. Widespread and common, Spindle-Shaped Coral fruits in clusters on the ground in woods. Edible, but tastes bitter.

Ramariopsis laeticolor

Fruitbodies are unbranched, up to 6 cm tall, flattened, often twisted, with a rounded apex, yellow to orange-yellow, not hollow, and mild-tasting. Widespread and common, this species fruits in scattered groups on the ground in woods. Also known as *Clavaria pulchra*.

Clavulinopsis corniculata

Fruitbodies are up to 5 cm tall by 2.5 cm wide, sparingly branched, and yellow to yellow-ochre. Stalks are whitish below to lemon-yellow above. Widespread and not uncommon, this species fruits on the ground in woodlands, in either solitary or scattered groups.

Clavicorona pyxidata

Crown Coral

Fruitbodies are branched, up to 12 cm tall and about as wide, pale yellow to pinkish-buff or tan, staining darker in age, and mild-tasting. Branching at right angles gives this species a candelabra-like appearance, with each branch terminating in a tiny, crown-like tip. Spore print is white. Common and widespread, Crown Coral fruits on well-rotted logs. Edible.

Ramaria aurea

Golden Coral

Fruitbodies are 7–15 cm tall by 10–15 cm wide, densely branched, golden-yellow to ochre, fading in age to drab brown, and with a short, thick, whitish stalk. Widespread and common, this species fruits on the ground in mixed woods. There are several similar species of yellow and orange-yellow coral fungi that are very difficult to tell apart. The edibility of many is not established, so **none** can be recommended.

Ramaria abietina

Fruitbodies are densely branched, up to 6 cm tall by 2–4 cm wide, yellowish-tan to light olive, with acutely pointed tips, and bitter-tasting. Coral fungi are difficult to identify. This species is confirmed microscopically by its very spiny spores (6–8 µm by 4–4.5 µm). Common and widespread, it fruits scattered or often in dense masses under conifers in summer.

Ramaria formosa

Pink-Tipped Coral

Fruitbodies are densely branched, 7–13 cm tall and almost as wide, and pinkish-buff or salmon-pink to orange-pink. Pinkish tinge disappears quickly, and the fruitbodies become ochre and stain brown with handling. Flesh in larger branches is slightly gelatinized. Spore print is yellow-brown. Widespread and locally common, Pink-Tipped Coral fruits on the ground in woods. **Poisonous.**

Ramaria botrytis

Clustered Coral

Fruitbodies are 7–15 cm tall by 3–15 cm wide, robust but brittle, pinkish-white to pink or purplish, bruising brown, and with vinaceous-brown tips. Fruitbodies have short, dense branches arising from a thick stalk to give a cauliflower-like appearance. Spore print is yellowish to orange-brown. Widespread and not uncommon, Clustered Coral fruits on the ground in woods. Edible.

Ramariopsis crocea

Orange-Yellow Ramariopsis

Fruitbodies are sparingly branched, 1–2 cm tall, deep orange or reddish-orange, fading to bright yellow, and with a delicate, fragile appearance. Scattered on bare ground or on humus, this delicate and beautiful species is widespread and not uncommon, but escapes attention because of its small size.

Ramariopsis kunzei

Fruitbodies are 2–6 cm tall, chalk-white, and with open branches arising from an inconspicuous stalk. Branches are slender, fragile, curved, and of uniform width. Common and widespread, this species is attractive and easily identified and fruits on the ground in grassy places in woods. Edible.

Ramaria stricta

Straight-Branched Coral

Fruitbodies are up to 8.5 cm tall by 20 cm wide, buff to ochre, erect, tightly branched, tough but not brittle, bruising vinaceous-brown, and with a bitter taste. Spore print is yellowish. Widespread and common, this densely branched coral fruits on well-rotted wood.

Lentaria byssiseda

Fruitbodies are up to 5 cm tall and pink to pale brown, staining darker. Branches are slender and more or less erect, and the tips remain pale. Spore-mass is cream to tan. Widespread and not uncommon, this delicate coral fungus fruits on small branches, twigs, or leaves on hardwoods.

Multiclavula mucida

Fruitbodies are unbranched or sometimes forked, less than 1 cm tall, very slender, often curved, with a few short laterals, and white to cream or yellowish. Fruitbodies taper towards the tips, which often turn brick-red to blackish, as if frosted. Clubs are pliable and persist for some time. Widespread and common, this tiny, lichenized coral fungus is always associated with green algal coatings on debarked logs.

Sparassis herbstii

Eastern Cauliflower

Fruitbodies are 5–30 cm or more wide, forming a dense clump of flat, wavy, noodle-like branches that are pallid to yellowish or pale ochre. Branches arise from a short, broad, rooting stalk. Widespread but not common, Eastern Cauliflower fruits at the base of trees. Edible.

Photo: Greg Thorn.

TOOTH FUNGI

Basidiomycota—Hymenomycetes

Tooth Fungi contain relatively few species of macrofungi. A few tooth fungi, however, are common, widespread, edible and easily recognized. In this group, the spores are produced on spore mother cells that form a layer (hymenium) on the outside of tooth-like spines. The spines develop on the underside of mushroom-like caps, as in Hedgehog Mushroom (*Hydnum repandum*, p. 123), or from fertile branches growing out from the sides of logs, as in Comb Tooth (*Hericium coralloides*, p. 122). In Icicle Fungus (*Mucronella bresadolae*, p. 127), spines hang like tiny stalactites from the sides or ends of well-rotted logs. Shelving Tooth (*Climacodon septentrionale*, p. 128) forms large, shelving brackets on the sides of hardwood trees, and some species, like *Steccherinum ochraceum* (p. 128), form flat (resupinate) growths over the surface of dead twigs or branches.

Most tooth fungi (e.g., *Hydnellum peckii*, p. 125) are tough, woody and inedible; a few (e.g., species of *Hydnum* and *Hericium*) are soft, fragile and highly prized as edibles. Tooth fungi have the advantage that they are easy to identify and the possibility of error is slim.

The common species have a widespread distribution across the country. As a group, however, the woody tooth fungi are much more common in the east than they are around the Great Lakes, and they are especially well represented in Nova Scotia, where the Canadian mycologist Ken Harrison discovered and described many species previously unknown. It is not easy to identify woody tooth fungi. For those who would like to try, many species are described in some detail in *How to Know the Non-Gilled Fleshy Fungi*.

Note on Edibility:

Do not eat **any** fungus in this group unless it is labelled edible. For further information, see pp. 317–18 for an illustrated list on recommended edible fungi.

Hericium coralloides

Comb Tooth

Fruitbodies are 5–20 cm wide and equally long, white, becoming cream to brown in age, and forming an open system of branches from which the fertile spines hang down. Spines are up to 10 mm tall and white to cream. Widespread and fairly common, Comb Tooth fruits on the side of dead hardwood trunks. Comb Tooth is distinguished from *H. americanum* (below) by its open branching and shorter spines, which give it an elegant, feathery appearance. This species is called *H. ramosum* in earlier books. Edible.

Hericium americanum

Fruitbodies are 5–20 cm wide by up to 25 cm long and consist of a thick support stalk, arising from the side of a log or stump, that produces a series of fertile branches. Each branch bears hanging clusters of spines, which are white at first, becoming creamy and finally brownish in age, and up to 40 mm long. Widespread and fairly common, this species fruits from the sides of hardwood logs. There has been considerable confusion in the literature in naming *H. coralloides*, *H. ramosum* (above), and *H. americanum*. This North American species is often referred to as *H. coralloides* in earlier books. Edible.

Hydnum repandum

Hedgehog Mushroom

Caps are 3–10 cm across, convex to flat, buff to orange-tan, smooth but often uneven, and with a wavy edge and low knob (umbo, see p. 181). Spines are up to 6 mm long, fragile, and cream to tan. Stalks are up to 8 cm tall by 2.5 cm wide and white to buff. Spore print is white. Widespread and fairly common, Hedgehog Mushroom fruits on the ground in mixed woods. Edible.

Hydnum umbilicatum

H. umbilicatum is very similar to Hedgehog Mushroom, but it is smaller and the cap seldom exceeds 5 cm across. Caps also have a deep depression at the centre, which often forms a hole (umbilicate, see p. 181). Widespread and fairly common, this species fruits in wet spots under conifers. Edible.

Bankera violascens

Caps are 3–12 cm across, convex to flat, depressed, and with a lobed or wavy margin. Surface is scaly, pinkish-brown to grey-brown, and with purplish tinges. Odour is pleasant. Teeth are up to 6 mm long and whitish to pale grey. Stalk is short and thick, up to 7 cm tall, tapering towards the base, and smooth to scaly. Widespread and not uncommon, this species fruits on the ground under conifers. Also known as *B. carnosa.*

Photo: Greg Thorn.

Sarcodon imbricatus

Caps are 5–20 cm across, convex to flattened, with a central depression, and covered with coarse, dark brown, upturned scales in a zonate arrangement on a pale brown background. Teeth are up to 1 cm long and whitish to tan, becoming brown with age. Stalks are up to 8 cm tall by 2.5 cm wide, central to off-centre, and hollow. Flesh is white, with a disagreeable taste. Widespread but not common, this species fruits under conifers.

Sarcodon scabrosus

Bitter Tooth

Caps are up to 20 cm across, convex, becoming flat, with an incurved margin, yellow-brown to pale red-brown, and covered with dark brown, adpressed scales. Taste is bitter. Teeth are up to 5 mm long. Stalks are up to 10 cm tall, with a pointed base. Widespread but not common, Bitter Tooth fruits on the ground under conifers.

Hydnellum spongiosipes

Spongy Foot

Caps are 2–10 cm across, convex to flat, azonate, brown to vinaceous-brown or rust-brown, and velvety to hairy. Teeth are crowded, up to 5 mm long, and buff to pinkish-brown, becoming cinnamon-brown but bruising dark brown. Stalks are central and up to 5 cm long and 1.5 cm wide at the top, but swell below to form a thick, spongy base up to 6 cm across. Widespread and fairly common, Spongy Foot fruits under hardwoods, especially oak.

Hydnellum geogenium

Yellow Tooth

Caps are up to 7 cm across, olive-black towards the centre, with a broad, tan to pale brown, irregular margin, weakly zonate, and often fused laterally. Teeth are up to 5 mm long, narrow, and sulphur-yellow at first, ageing to brown. Stalk is absent and fruitbodies are attached to the substrate by bright yellow strands. Locally common, Yellow Tooth fruits under spruce and fir in coastal areas.

Hydnellum peckii

Bleeding Tooth

Caps are 5–15 cm across, flat to slightly depressed, felty to woolly, dark brown at the centre, changing to salmon-pink at the margin, and exuding blood-red droplets when young. Teeth are salmon-pink, ageing to brown, and up to 4 mm long. Stalks are up to 8 cm tall by 3 cm wide and velvety. Flesh is dingy-brown, with a peppery taste. Widespread and not uncommon, Bleeding Tooth fruits on the ground under conifers.

Hydnellum caeruleum

Blue Tooth

Caps are 3–11 cm across, violet-blue when young, fading to white, and ageing to dark brown. Surface is velvety, becoming pitted in age. Teeth are bluish, becoming whitish, and then dark brown, with whitish tips. Stalks are up to 4 cm tall by 2 cm wide. Widespread but not common, Blue Tooth fruits under conifers in northeastern North America.

Hydnellum concrescens

Zonate Tooth

Caps are up to 10 cm across, solitary or fusing at the margin, more or less smooth, and with a zonate surface and radial ridges. Caps are brown to purplish-brown and paler towards the margin. Teeth are up to 2 mm long and dark brown. Stalks are up to 4 cm long by 2 cm wide and coloured as the cap. Widespread and fairly common, Zonate Tooth fruits on the ground in hardwood forests. Also known as *H. zonatum*.

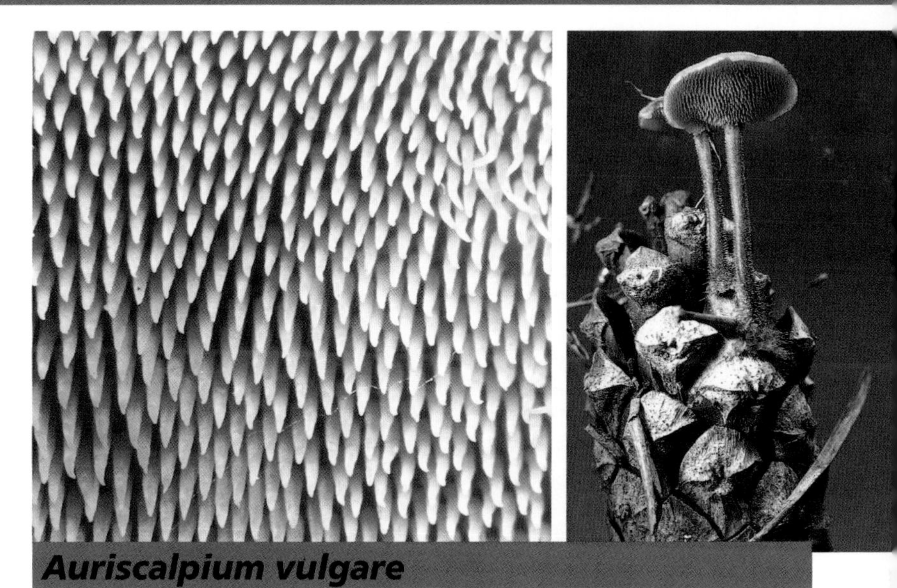

Auriscalpium vulgare

Pine Cone Fungus

Caps are 5–20 mm across, convex to flattened, dark brown, and circular to irregular in outline. Stalks are lateral, up to 7 cm tall by 2 mm across, tough, and densely covered with dark brown hairs. Teeth are white to violet-brown and up to 2.5 mm long. Widespread and common, Pine Cone Fungus fruits on the cones of pine and fir.

Mucronella bresadolae

Icicle Fungus

Fruitbodies form icicle-like spines, solitary or in clusters, arising directly from the substrate. Spines are white, up to 3 mm long, and tapering to an acute point. Widespread but not common, Icicle Fungus fruits on well-rotted wood. In *M. pendula* the spines are larger and solitary.

Climacodon septentrionale

Shelving Tooth

Fruitbodies are in tight, shelving layers. Brackets are broadly attached, 10–28 cm across by up to 12 cm wide and 5 cm thick at the base, yellowish-white, becoming tan with age, and covered with densely matted hairs. Flesh is thick, white, tough, and fibrous. Spines are up to 2 cm long, tightly packed, and whitish, ageing to tan. Widespread and not uncommon, Shelving Tooth fruits on the trunks of living hardwood trees, especially maples. Also known as *Steccherinum septentrionale*.

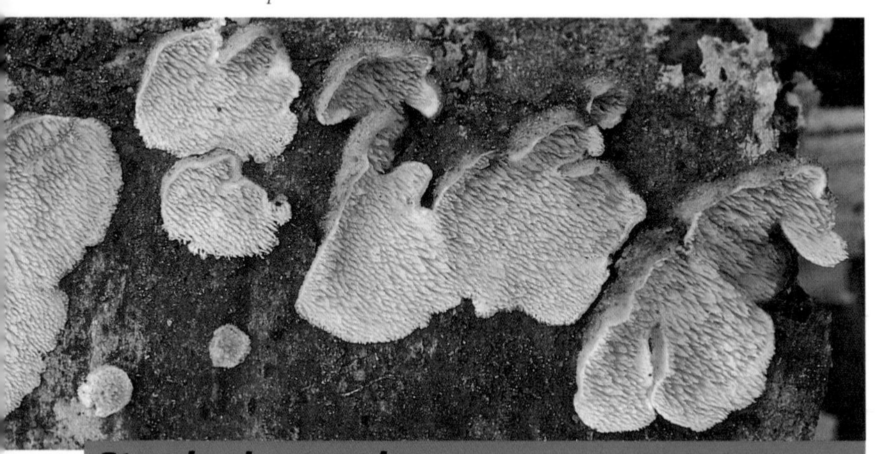

Steccherinum ochraceum

Fruitbodies are flat on the surface (resupinate) or sometimes in shelving brackets, very variable in size, 3–30 cm long, 1–3 mm thick, and ochre. When present, brackets are up to 4 cm wide, with a hairy, zonate surface, and often laterally fused. Spines are up to 1.5 mm long and ochre. Widespread and not uncommon, this species fruits on dead hardwood twigs and branches.

BRACKET FUNGI

Basidiomycota—Hymenomycetes

In **Bracket Fungi** the spores are produced inside tubes that line the underside of the fruitbody. These tubes open by pores to the exterior, giving a perforated appearance to the undersurface (Fig. 8). For this reason, species in this group are sometimes called **polypores.** Tubes can be shallow or more than a centimetre deep. Pores are often **circular** in the face view, but might be **angular** (p. 133), **elongate** (p. 142) or even **labyrinthiform** (maze-like, p. 148). In a few cases, the walls between pores break down, to give a gill-like appearance to the underside of the fruitbody, as in *Lenzites* spp. (p. 143) or in older specimens of *Daedaleopsis confragosa* (p. 143). Pores range in size from almost a centimetre across to a fraction of a millimetre. In Artist's Conk (*Ganoderma applanatum*, p. 139), they are so narrow that they are difficult to see without a magnifying glass. Deeper tubes and a smaller pore size both serve to increase the spore-producing surface (hymenium).

For the most part, fruitbodies of bracket fungi are tough and leathery or woody in texture. Inside, there are thin-walled, living hyphae (filaments) for the transport of nutrients and production of spores. As well, most bracket fungi contain thick-walled, dead hyphae. These thick-walled, branching fibres interlock and form an extremely hard and rigid fruitbody that can persist for long periods, sometimes for many years. The hardness of the fruitbody will be related to the proportion of these thick-walled cells.

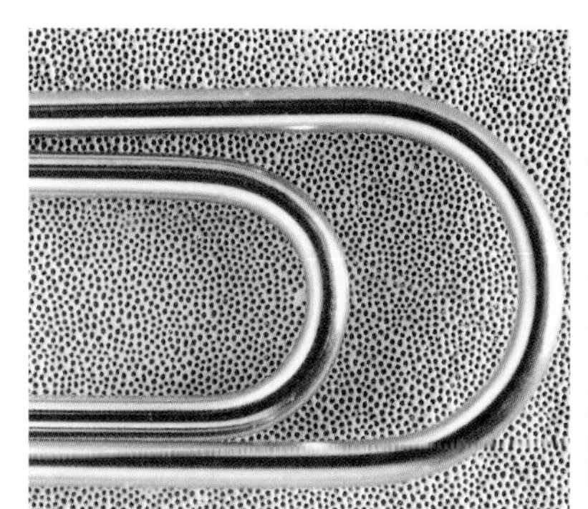

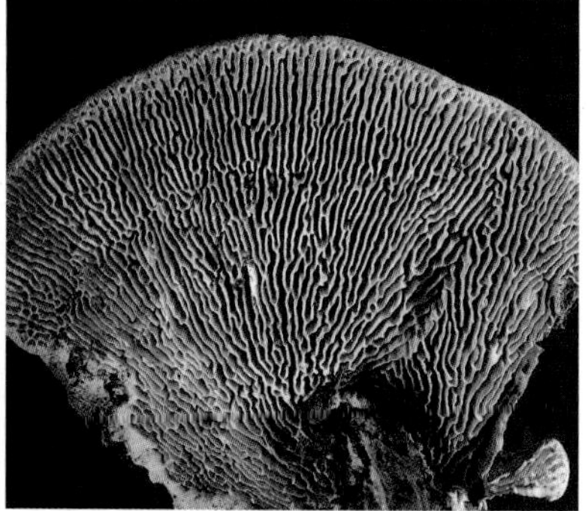

Fig. 8. **Pore Types in Bracket Fungi. Left:** tiny, circular pores of *Ganoderma applanatum*. Standard paper clip gives scale. **Right:** Labyrinthiform (maze-like) pores of *Daedalea quercina*.

Bracket fungi are so-called because they grow from the sides of trees, much like shelves. Some are large and robust, whilst others are thin and delicate. A few polypores do not form brackets and are mushroom-like, with a cap and a central stalk. (For more information on mushroom cap shapes, see p. 181.) At first glance, they look like boletes, but their tough, leathery texture gives them away. Also, in some polypores, shelves are lacking and fruitbodies form a flat layer (resupinate) on the undersurface of a twig or branch.

Most mushrooms turn into a putrefying mass within days. Bracket fungi, however, can last for weeks to months or even overwinter. A few, such as Artist's Conk, can persist for many years, producing a new layer of pores on the underside each year. During summer heat or drought, brackets might shrivel a bit, but can recover in rainy periods to resume spore production.

Bracket fungi are not the most attractive fungi and tend to be drab and uninteresting. There are exceptions to any rule, however, and several bracket fungi are colourful. Chicken of the Woods (*Laetiporus sulphureus*, p. 149) is rich yellow to orange-yellow, with reddish tints. Cinnabar Polypore (*Pycnoporus cinnabarinus*, p. 149) is orange to faded orange on the upper surface and brilliant cinnabar underneath. Turkey Tail (*Trametes versicolor*, p. 138) has attractive zonation, ranging from tans to rich browns in a velvety textured surface. Attractive honey-coloured resin droplets exude from the edge of Late Fall Polypore (*Ischnoderma resinosum*, p. 140), and Red-Banded Polypore (*Fomitopsis pinicola*, p. 142) has a colourful orange-brown band lining the margin of the cap.

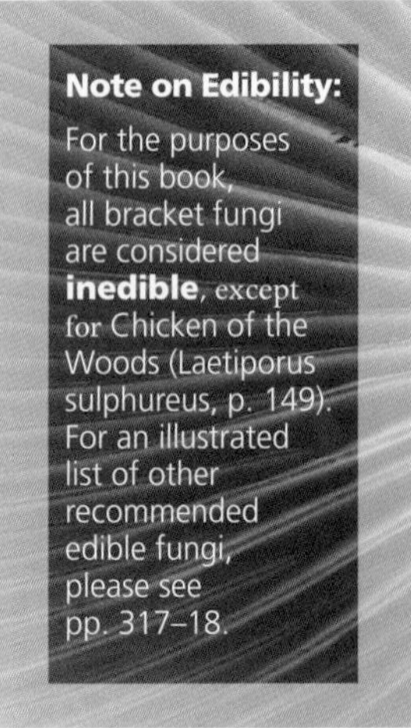
Note on Edibility:

For the purposes of this book, all bracket fungi are considered **inedible**, except for Chicken of the Woods (Laetiporus sulphureus, p. 149). For an illustrated list of other recommended edible fungi, please see pp. 317–18.

Bracket fungi are wood rotters *par excellence* and, like many other fungi, play a vital role in the carbon cycle in processing woody debris (see p. 26 for further information). It is common, therefore, to find bracket fungi fruiting on logs, stumps, branches or twigs. Some bracket fungi are parasites and grow on living trees. A few, such as Rooting Polypore (*Polyporus radicatus*, p. 147), colonize buried wood and produce stalked fruitbodies, seemingly directly from the ground.

Because of their tough, woody consistency, bracket fungi are not highly regarded as edibles. Nevertheless, a few have made the gourmet's list. Amongst these bracket fungi is the aforementioned Chicken of the Woods. The young margins (edges) of the expanding brackets are reported to be edible and very good, but this high opinion is by no means unanimous.

Some bracket-like fungi—stereoid fungi—have no pores. The underside might be uneven, but it is more or less smooth. Sometimes stereoid fungi look like paint on wood, or they form shelving brackets and mimic the true bracket fungi. *Stereum ostrea* (p. 154) is called False Turkey Tail because of its superficial resemblance to Turkey Tail, a true bracket fungus; the two are easily separated by looking at the undersides. Turkey Tail has pores, and False Turkey Tail has none.

Key to Genera of Bracket Fungi

FB = Fruitbodies

1. FB with pores, which can split to form teeth or gills **2**
1. FB without pores and smooth to wrinkled or bumpy or flat (resupinate) **33**

2. Fruiting on the ground **3**
2. Fruiting on wood **11**

3. Flesh < 1 cm thick .... ***Coltricia***
3. Flesh > 1 cm thick **4**

4. FB very large, composed of many separate caps on a common stalk **5**
4. FB with a single cap, although cap margins might fuse laterally **6**

5. Each cap < 5 cm in diameter .... ***Polyporus umbellatus***
5. Each cap > 5 cm in diameter, stalks bleed latex when cut .... ***Bondarzewia***

6. FB anchored by a long root, pores whitish .... ***Polyporus radicatus***
6. FB lacking root **7**

7. Caps blackish-brown to black, pores whitish .... ***Boletopsis***
7. Caps whitish, yellow-brown, or red-brown **8**

8. FB clustered, often fused at the margins .... ***Albatrellus***
8. FB solitary or several but never fused **9**

9. FB grey-brown to red-brown **10**
9. FB yellow-brown .... ***Albatrellus, Inonotus***

10. Pore surface white .... ***Polyporus***
10. Pore surface yellow, becoming brown .... ***Phaeolus***

11. FB with a central stalk .... ***Polyporus***
11. Stalks lateral or absent **12**

12. FB yellow to orange or cinnabar **13**
12. FB other colours **15**

13. FB orange above, bright cinnabar below .... ***Pycnoporus***
13. FB yellow to yellowish to orange-yellow **14**

14. FB large (up to 30 cm wide), flesh thick, shelving .... ***Laetiporus***
14. FB smaller, with large, angular pores .... ***Polyporus, Pycnoporellus***

15. FB white to cream or bluish **16**
15. FB grey, brown to blackish (often green with algae) **18**

16. FB thin and tough, < 1 cm deep .... ***Trametes***
16. FB soft to spongy, > 1 cm deep **17**

17. FB broader than deep, with one pore layer .... ***Postia, Tyromyces***
17. FB deeper than broad, with several pore layers .... ***Oligoporus***

18. Pores labyrinthiform (maze-like) or gill-like **19**
18. Pores round to angular **23**

19. FB thick, up to 8 cm .... ***Daedalea***
19. FB < 3 cm thick **20**

20. Pore surface gill-like .... ***Lenzites***
20. Pore surface labyrinthiform **21**

21. Surface of FB smooth .... ***Daedaleopsis***
21. Surface of FB densely hairy **22**

22. FB red-brown, on conifer wood .... ***Gloeophyllum***
22. FB whitish to pale brown, turning green, on hardwood .... ***Cerrena***

23. FB < 1 cm thick **24**
23. FB > 1 cm thick **26**

24. Pore surface smoky-brown to blackish-grey .... ***Bjerkandera***
24. Pore surface white to cream or purple-tinged **25**

25. Pore surface white to cream .... ***Trametes***
25. Pore surface purple, especially towards the margin .... ***Trichaptum***

26. FB perennial (tubes in layers) **27**
26. FB annual (never more than one layer of tubes) **31**

27 FB with bright red-brown band near the perimeter .... ***Fomitopsis***
27. FB lacking red band **28**

28. FB about as deep as wide **29**
28. FB much wider than deep **30**

29. FB hoof-shaped, pore surface exposed .... ***Fomes***
29. Pore surface covered by membranous sheath .... ***Cryptoporus***

30. FB with a black, charred surface .... ***Phellinus***
30. FB brown to blackish or lacquered .... ***Heterobasidion, Ganoderma***

31. FB grey, corky, on birch .... ***Pycnoporus***
31. FB yellow-brown or deep red-brown **32**

32. FB yellow-brown .... ***Hapalopilus***
32. FB deep red-brown .... ***Ischnoderma***

33. FB on the ground **34**
33. FB on wood **36**

34. FB stalked **35**
34. FB not stalked .... ***Thelephora***

35. FB white to cream .... ***Cotylidia***
35. FB brown, zonate .... ***Thelephora***

36. FB bracket-form in overlapping shelves **37**
36. FB flat on surface of wood **39**

37. FB soft, gelatinous .... ***Phlebia***
37. FB tough **38**

38. Underside rust-brown to dark brown .... ***Hymenochaete***
38. Underside dirty white to pale brown .... ***Plicaturopsis, Stereum***

39. FB in tile-like mosaic .... ***Xylobolus***
39. FB not as above **40**

40. FB forming smooth or wrinkled pads over wood .... ***Peniophora***
40. FB forming a radially wrinkled, pinkish growth over wood .... ***Phlebia***

Albatrellus ovinus

Sheep Polypore

Fruitbodies are stalked. Caps are 5–15 cm across, irregularly circular, flat to depressed, dry, smooth or irregularly cracked, at first soft and fleshy, becoming tough in age, and off-white to yellowish, becoming tan to smoky. Tubes are shallow. Pores are white to yellowish and 3–5 per mm. Stalks are whitish, more or less central, and up to 8 cm tall by 3.5 cm wide. Widespread and fairly common, Sheep Polypore fruits under conifers.

Photo: Greg Thorn.

Albatrellus confluens

Fruitbodies are stalked. Caps are up to 12 cm across and 3 cm thick, flat or somewhat depressed towards the centre, more or less circular, with an even or lobed margin, fleshy to soft, and cream to tan or pinkish-buff to light orange-brown. Fruitbodies are often fused together at the cap margins. Pores are white to cream. Stalks are up to 8 cm long by 2 cm wide, tapering towards the base, and white to cream. Widespread and not uncommon, this species fruits under conifers.

Albatrellus caeruleoporus

Blue Albatrellus

Fruitbodies are stalked. Caps are 2–6 cm across, often fused, convex, becoming flat, blue to blue-grey, and browning in age. Tubes are up to 3 mm long, 2–3 per mm, and with angular pores. Stalks are up to 7 cm tall by 2.5 cm wide and central to off-centre. Widespread but not common, Blue Albatrellus fruits under hemlock.

Photo: Greg Thorn.

Bondarzewia berkeleyi

Fruitbodies are 25–90 cm wide and up to 30 cm tall, and with one, several, or many caps arising from a single root-like base. Caps are pale buff to tan or greyish-brown, spongy-tough, and with a radially roughened surface. Pore surface is whitish, with circular to angular or sometimes labyrinthiform pores. If cut when fresh, the pores exude latex. Widespread but not common, this species causes butt rot of standing trees and fruits on the ground close to the trunks of living hardwoods, particularly oak.

Polyporus umbellatus

Fruitbodies are stalked, up to 50 cm across, and composed of numerous more or less circular, centrally attached caps, arising from a common, strongly branched stalk. Individual caps are up to 3 cm across, flat, thin, smooth, and ochre to pale brown. Pores are angular, 1–3 per mm, and whitish to straw-coloured. Stalks are up to 3 cm wide. Widespread but not common, this species fruits on the ground near hardwood stumps. Also known as *Grifola umbellata*.

Photo: Greg Thorn.

Coltricia perennis

Fruitbodies are stalked. Caps arc 2–10 cm wide, up to 5 mm thick, flat to depressed or funnel-shaped, and velvety. Caps are concentrically zoned into a series of golden-yellow to cinnamon-brown bands. Pores are angular, 2–4 per mm, golden-brown to dark brown, and with shallow tubes. Stalks are up to 5 cm tall by 1 cm wide, hairy to velvety, and coloured as the cap. Widespread and common, this species fruits on the ground in sandy soil under conifers, often at the edge of trails and woodland paths.

Coltricia cinnamomea

Fairy Stool

Fruitbodies are stalked. Caps are circular, up to 4 cm across, flat to funnel-shaped, silky to finely hairy, sometimes shiny, very thin, zonate, and brown to deep red-brown. Stalks are up to 4 cm tall by 4 mm wide and red-brown. Pores are angular, red-brown, and 2–3 per mm, with shallow tubes. Widespread and fairly common, Fairy Stool fruits on the ground in deciduous woods, often beside trails.

Coltricia montagnei

Fruitbodies are stalked. Caps are up to 12 cm across, circular, convex to flat, becoming depressed at the centre, cinnamon to rust-brown or dark brown, hairy, and weakly zonate. Stalks are up to 4 cm tall by 1 cm wide and cinnamon to rust-brown. Tubes are up to 5 mm deep, with angular pores, sometimes becoming gill-like. Widespread but not common, this species fruits on the ground in hardwood forests.

Photo: Brian Shelton.

Postia caesia

Fruitbodies are shelving, 3–6 cm long by 1–4 cm wide and up to 1 cm thick, white to off-white, soft to spongy, often stained bluish-grey, velvety to finely hairy, becoming smooth in age, and azonate. Flesh is white, 1–10 mm thick, and mild-tasting. Pores are white to grey or blue-grey. Widespread and common on dead wood, this species is also known as *Tyromyces caesius* and *Oligoporus caesius*.

Postia fragilis

Fruitbodies are bracket-form, semi-circular to elongate, 2–6 cm long by 4 cm wide and 2 cm thick, and often fusing laterally, bruising brown. Pores are angular, 2–3 per mm, and white, rapidly bruising brown when handled. Widespread and not uncommon, this species fruits on coniferous wood. Also known as *Tyromyces fragilis* and *Oligoporus fragilis*.

Tyromyces chioneus

Fruitbodies are bracket-form, convex to flat, up to 10 cm long by 8 cm wide and 0.5–3 cm thick, soft, pure white, finely hairy, becoming smooth in age, and azonate. Flesh is white. Pores are white, angular, and 3–4 per mm. Widespread and common on dead wood, this species is also known as *T. albellus*.

Photo: Greg Thorn.

Trametes conchifer

Fruitbodies are shell-like in early stages, up to 2 cm across, later cup-shaped, zonate, and in shades of grey, white, and brown. The growth continues from the cups to produce shelf-like brackets up to 5 cm long by 3 cm wide and 3 mm thick, whitish to buff or yellowish-tan, zonate, and smooth to slightly velvety. Pores are angular and 2–4 per mm. Also known as *Poronidulus conchifer*. Widespread, this species is found on dead hardwood twigs and branches.

Trametes hirsuta

Fruitbodies are shelving, broadly attached, semi-circular, up to 10 cm long by 6 cm wide, up to 2 cm thick, tough, flexible, white to grey, becoming dirty brown, densely hairy, and azonate to weakly zonate. Pores are white to cream, becoming straw-coloured in age, and 3–4 per mm. Also known as *Coriolus hirsutus*. Widespread and not uncommon, this species fruits on hardwood logs or stumps.

Trametes pubescens

Fruitbodies are shelving, overlapping, broadly attached, up to 6 cm long by 5 cm wide, less than 1 cm thick, cream to yellowish-tan, weakly zonate, and tough. Upper surface is smooth to finely hairy. Pores are white, ageing yellowish, and 3–4 per mm. Common and widespread, this species fruits on dead hardwood branches.

Trametes versicolor

Turkey Tail

Fruitbodies are shelving, overlapping, thin, tough, 3–8 cm long by up to 5 cm wide, 1–3 mm thick, velvety, narrowly zonate, and variable in colour, from tan to orange to red-brown to amber. Pores are white to yellowish and 3–5 per mm, with shallow tubes. Widespread and very common, Turkey Tail fruits on hardwood logs and stumps. Also known as *Coriolus versicolor.*

Ganoderma applanatum

Artist's Conk

Fruitbodies are bracket-form, broadly attached, up to 50 cm long by 30 cm wide, woody, smooth, and concentrically grooved in grey-brown to dark brown zones. Pores are white to cream, bruising dark brown, narrow, and 4–6 per mm, in layers (perennial). Widespread and very common, Artist's Conk fruits on living trees or recently cut hardwood stumps and logs.

Ganoderma lucidum

Lacquered Polypore

Fruitbodies are 5–30 cm wide, bracket-form or stalked, semi-circular to fan-shaped, smooth, azonate to concentrically grooved, yellowish near the margin, red-brown to mahogany in older parts, and with a lacquered surface. Pores are white to cream, ochre in age, circular, and 4–6 per mm in a single layer (annual). Widespread and fairly common, Lacquered Polypore fruits on hardwoods. *G. tsugae*, a similar species, fruits on conifers.

Ischnoderma resinosum

Late Fall Polypore

Fruitbodies are bracket-form, 7–25 cm long by 3–15 cm wide, dark brown, velvety, becoming radially furrowed, and exuding amber-coloured fluid when young. Flesh is spongy, tough, watery, and straw-coloured. Tubes are white, staining brown. Pores are round to angular and 4–6 per mm. Widespread and common, Late Fall Polypore fruits on hardwood logs and stumps late in the year.

Heterobasidion annosum

Fruitbodies are flat to irregular or shelving, up to 20 cm across by 9 cm wide and 5 cm thick, hard, woody, hairy, becoming smooth, with an uneven surface and irregular concentric grooves, light brown to red-brown or dark brown, with a white margin, and turning blackish in age. Pores are ivory to cream, round to angular, and 4–5 per mm. Widespread and not uncommon, this species fruits on conifer logs and stumps or is parasitic on the roots of conifers.

Phellinus igniarius

Fruitbodies are 25 cm long by 15 cm wide and 12 cm thick, bracket-form, broadly attached, semi-circular to elongate, convex to flat, hard, woody, smooth at first, and with concentric grooves. Upper surface becomes black and fissured to give a charred appearance. Pores are 5–6 per mm, brown, and in layers (perennial). Widespread and not uncommon, this species fruits on hardwood logs and stumps.

Fomes fomentarius

Tinder Polypore

Fruitbodies are up to 15 cm wide, bracket-form, broadly attached, woody, hoof-shaped, with the height often exceeding the width, smooth, zonate, and light grey to blackish-grey and shades of brown. Pores are beige to ochre, becoming dark brown, round, 4–5 per mm, up to 1 cm thick, and in layers (perennial). Widespread and very common, Tinder Polypore fruits on dead deciduous trees and logs.

Cryptoporus volvatus

Fruitbodies are up to 5 cm across by 4 cm wide and 4 cm deep, top-shaped to turnip-shaped, laterally attached, tan-coloured, often appearing to be lacquered, corky to woody, and with a tough membrane forming a chamber below the pore surface. Membrane eventually opens by a large, circular hole. Pores inside chamber are tiny and white to brown. Widespread but infrequent in Ontario, eastern Canada and the adjacent U.S., this distinctive species fruits prolifically on recently dead standing conifers.

Fomitopsis pinicola

Red-Banded Polypore

Fruitbodies are bracket-form, broadly attached, woody, up to 40 cm long by 25 cm wide and 15 cm thick, flat, concentrically grooved, yellowish to orange-red, becoming red-brown and finally blackish, and with a resin-coated surface. Distinct red-brown band runs along the upper surface near the margin of the bracket. Pores are cream-coloured to light brown or ochre and 5–6 per mm in layers (perennial). Widespread and common, Red-Banded Polypore fruits on logs and stumps, especially spruce.

Fomitopsis cajanderi

Fruitbodies are bracket-form or shelving, broadly attached, flat, woody, up to 20 cm across the base by 10 cm wide and 7 cm thick, tan, with a purplish tinge to pinkish-brown, and becoming blackish in age. Flesh is spongy and pinkish. Pores are pinkish-brown, becoming smoky-brown, and 4–5 per mm. Widespread but not common, this species fruits on conifer wood.

Gloeophyllum sepiarium

Fruitbodies are bracket-form, shelving, up to 12 cm long by 7 cm wide and 1 cm thick, broadly attached, semi-circular to discoid, orange-brown to rusty or dark brown, zonate, and densely hairy. Pores are elongate to labyrinthiform, or often gill-like, and yellow-brown to red-brown or dark brown. Widespread and common, this species fruits on exposed coniferous wood, such as docks, railway ties, and slash.

Daedaleopsis confragosa

Fruitbodies are shelving, up to 12 cm wide by 2 cm thick, semi-circular, broadly attached, tan to brown, often with a pinkish tinge, smooth, zonate, corky to tough, concentrically furrowed, and pale brown to dark brown. Pores are circular to elongate or labyrinthiform to gill-like, white to buff, bruising pink, and ageing brown. Widespread and very common, this species fruits on hardwood branches.

Daedalea quercina

Oak Polypore

Fruitbodies are bracket-form, broadly attached, up to 20 cm long by 15 cm wide and 8 cm thick, corky to woody, flat, zonate, with irregularly spaced and raised ridges, smooth, and tan to brown, becoming dark brown with age. Pores are labyrinthiform to gill-like, 1–4 mm wide, pale brown, and thick-walled. Wide-spread and common, Oak Polypore fruits on logs or stumps of oak.

Lenzites betulina

Birch Lenzites or Gilled Bracket

Fruitbodies are bracket-form, up to 8 cm long by 5 cm wide by 2 cm thick, semi-circular, thin, tough, hairy, radially grooved, with concentric bands of colour, brown to orange-brown or rusty, and with the underside white to cream or ochre. Pores are lacking and are replaced by gills radiating from the attachment point. Widespread and not uncommon, Birch Lenzites fruits on hardwood stumps and logs.

Polyporus mori

Fruitbodies are stalked. Caps are 8 cm across, 7 mm thick, kidney-shaped to fan-shaped, with a wavy or scalloped margin, yellowish to brick-red, and smooth to fibrillose-streaked or scaly. Pores are 1–3 mm across, white to cream, diamond-shaped, large, and radially arranged. Stalks are lateral to central, short, and stout. Widespread and common, this species fruits early in the year on dead branches of deciduous trees. Also known as *P. alveolaris* and *Favolus mori*.

Polyporus arcularius

Spring Polypore

Fruitbodies are stalked. Caps are up to 4 cm across, dry, convex to umbilicate or funnel-shaped, with a hairy margin, yellow-brown to dark-brown, thin, and tough. Pores are white to yellowish, angular, and up to 2 mm across. Stalks are central, up to 6 cm tall by 4 mm wide, ochre to dark brown, and hairy or smooth. Widespread and not uncommon, Spring Polypore fruits on dead hardwood in spring and early summer.

Phaeolus schweinitzii

Dye Maker's Polypore

Fruitbodies are more or less circular, up to 25 cm across by 3 cm thick, flat or depressed, corky-tough, finely or coarsely hairy, becoming smooth in age, golden to ochre or rusty to red-brown, and with a yellowish margin. Pores are yellow, becoming brown, and 1–3 per mm. Tubes are 1–4 mm deep. Widespread and fairly common, Dye Maker's Polypore fruits on the ground from buried roots or on stumps of conifers.

Polyporus squamosus

Dryad's Saddle

Fruitbodies are bracket-form to short-stalked, up to 30 cm across by 5 cm thick, kidney-shaped to nearly circular, dry, yellowish to tan, covered with dark brown, flattened scales, and fleshy, becoming tough. Pores are large, angular, 1–2.5 mm wide, and off-white to yellowish. Stalks are lateral and up to 5 cm long by 3 cm wide. Widespread and common, Dryad's Saddle fruits in fall and spring on hardwood stumps and logs or living trees.

Photo: Greg Thorn.

Polyporus hirtus

Fruitbodies are centrally to laterally stalked. Caps are circular, up to 15 cm across, greyish to purple-brown, and hairy to scaly and flaky, becoming smooth. Pores are white to cream, angular, and 1–2 per mm. Stalks are up to 10 cm tall by 4 cm across, tan to purple-brown, and hairy to smooth. Widespread but not common, this species fruits on the ground under conifers. Also known as *Jahnoporus hirtus*.

Polyporus brumalis

Winter Polypore

Fruitbodies are stalked, 2–8 cm across, circular, flat to umbilicate, thin, with an inrolled margin, grey to umber to blackish, smooth, and azonate. Pores are angular, large, 0.5–1 mm wide, shallow, and with a white to cream surface. Stalks are up to 6 cm tall by 6 mm wide and light brown. Widespread and common, Winter Polypore fruits on dead hardwood logs in late fall.

Polyporus varius

Blackfoot Polypore

Fruitbodies are stalked. Caps are up to 8 cm across, circular to fan-shaped, flat to depressed, tough, and tan. Pores are angular, 4–5 per mm, shallow, and white to cream, becoming brown. Stalks are up to 4 cm long by 5 mm wide, smooth, tan, and becoming black, especially near the base. Widespread and common, Blackfoot Polypore fruits on dead twigs or branches of hardwoods.

Polyporus badius

Bay-Brown Polypore

Fruitbodies are stalked. Caps are up to 15 cm wide, flat to depressed, with a wavy edge, circular to kidney-shaped, bay-brown, fading to cream, especially at the margin, smooth, and glossy to dull. Pores are tiny, 6–9 per mm, and with a white to cream or ochre surface. Stalks are up to 8 cm tall by 1 cm wide and dark brown to black at maturity. Widespread and common, Bay-Brown Polypore fruits on dead hardwood logs or stumps.

Polyporus radicatus

Rooting Polypore

Fruitbodies are stalked. Caps are circular, up to 20 cm wide by 1 cm thick, convex to depressed, yellow-brown to smoky-brown, fleshy-tough, and scurvy to fibrillose or scaly, becoming smooth at the centre. Flesh is white and 3–10 mm thick. Stalks are long, black, scurfy, and rooting. Widespread and not uncommon, Rooting Polypore fruits on the ground near stumps in summer.

Boletopsis subsquamosa

Fruitbodies are stalked. Caps are up to 15 cm across by 4 cm thick, dirty brown to blackish, and with tough flesh. Margin is scalloped and wavy. Pores are angular, dirty white to pale grey, and 1–3 per mm. Stalks are central or lateral, not well differentiated from the cap, grey to brownish, and up to 7 cm tall. Spore print is whitish to tan. Widespread but not common, this species fruits on the ground, mostly under pines. As the name suggests, *B. subsquamosa* has the look of a bolete, but its tough texture is diagnostic. Also known as *B. leucomelas*.

Plicaturopsis crispa

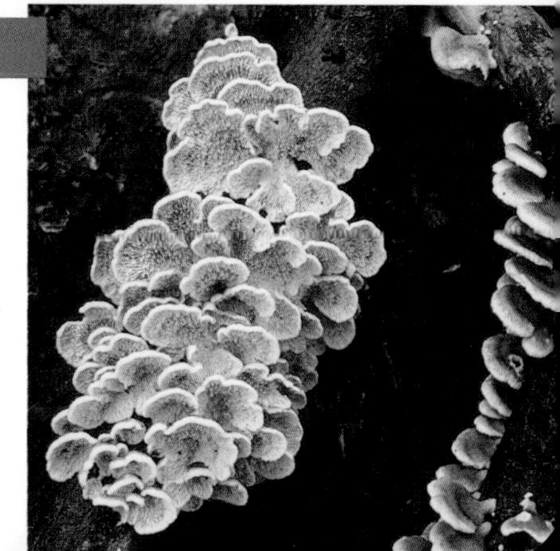

Fruitbodies are shelving or with short stalks, 10–20 mm wide, thin, tough, ochre to red-brown, paler towards the margin, and with closely adpressed hairs over the surface. Fruitbodies shrivel when dry and revive during wet periods. Gills are white to pale grey, shallow, often forked or vein-like, crinkled, and irregular. Common and widespread, this species fruits on dead branches of hardwoods. This species is not a bracket fungus, but can be mistaken for one.

Cerrena unicolor

Fruitbodies are shelving, overlapping, 2–10 cm long by up to 6 cm wide, flat, densely hairy, concentrically zonate, thin, white to grey or pale brown, and frequently turning green because of algal growth. Flesh is thin and separated from the surface by a thin, black zone (seen on cutting). Pores are white to ochre, elongate to labyrinthiform, and sometimes tooth-like. Widespread and very common, this species fruits on hardwood stumps and logs.

Trichaptum biforme

Purple-Toothed Polypore

Fruitbodies are shelving and overlapping, flat to slightly convex, 2–8 cm long by 2–6 cm wide, thin, tough, velvety, narrowly zonate, and white to cream or ochre, becoming greyish in age. Pores are violet to bluish, especially near the margin, becoming brown, angular, 2–4 per mm, and tooth-like in age. Widespread and very common, Purple-Toothed Polypore fruits on deciduous wood and is also known under *Hirschioporus biformis*.

Trichaptum abietinum

Similar in general appearance to *T. biforme* (above), but this species is smaller and thinner, and it fruits on conifers. Widespread and common, this species is also known as *Hirschioporus abietinus*.

Pycnoporus cinnabarinus

Cinnabar Polypore

Fruitbodies are bracket-form, up to 10 cm long by 8 cm wide and 2 cm deep, semi-circular, more or less flat, corky to woody, smooth, azonate, and pale orange to cinnabar, ageing whitish. Pores are angular, 2–3 per mm, up to 7 mm deep, and brilliant orange-red. Widespread but not common, Cinnabar Polypore fruits on hardwood logs.

Piptoporus betulinus

Birch Polypore

Fruitbodies are bracket-form, solitary, up to 25 cm at their longest by 15 cm wide and 6 cm thick, dirty white to grey or brownish, smooth or with irregular markings, sometimes becoming scaly-rough, and with a soft to corky, light-weight interior. Pores are smooth at first, spherical to angular, 3–5 per mm, drying and splitting to become uneven in age, and with a tooth-like appearance. Stalks are short and stout. Widespread and very common, Birch Polypore fruits on dead birch.

Laetiporus sulphureus

Chicken of the Woods

Fruitbodies are bracket-form, up to 30 cm across, broadly attached, bright sulphur-yellow to yellow-orange, sometimes with a reddish tinge, smooth but uneven, and azonate. Pores are yellow, round, and 3–5 per mm. Widespread and relatively common, Chicken of the Woods fruits on both hardwood or coniferous stumps or living trees. **Margins** of brackets are edible.

Inonotus radiatus

Fruitbodies are shelving and overlapping, often fusing laterally to form an elongate series (of fruitbodies), ochre to rusty-brown, becoming black in age, zonate, silky to hairy, radially fibrillose, often grooved at the margin, and up to 7 cm along the base by 6 cm wide and 3 cm thick. Pores are angular, white, staining brown, and 4–5 per mm. Widespread and fairly common, this species fruits on hardwood logs.

Photo: Greg Thorn.

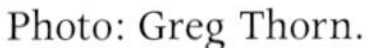

Inonotus tomentosus

Fruitbodies are centrally or laterally stalked. Caps are up to 10 cm wide, circular to semicircular, sometimes lobed, hairy, yellow-brown, and sometimes faintly zonate. Pores are angular and 2–4 per mm. Stalks are up to 4 cm tall by 1.5 cm wide. This species fruits on the ground under conifers or on the trunks of living or dead conifers, especially spruce. Also known as *Onnia tomentosa*.

Photo: Greg Thorn.

Hapalopilus nidulans

Fruitbodies are shelving, convex to flat, broadly attached, up to 10 cm wide by 4 cm thick at the base, tan-brown to ochre, spongy-tough, drying light, and brittle. Pores are angular, 2–4 per mm, and yellow-brown to cinnamon-brown. Spore print is white. This species fruits on dead hardwood logs or trunks.

Bjerkandera adusta

Fruitbodies are shelving, overlapping, narrow, 3–7 cm long, up to 4 cm wide, thin, pliable, velvety, becoming smooth, dirty white to grey or smoky-brown, weakly zonate or azonate, and becoming blackened in age, especially at the margin. Pores are grey to black, round to angular, and 6–7 per mm. Widespread and very common, this species fruits on hardwood stumps and logs.

Pycnoporellus alboluteus

Fruitbodies consist of cushion-shaped masses of pores on thin, orange flesh. Fruitbodies are bright orange, but might fade to yellowish-white. Pores are large, angular, at irregular levels, elongate, becoming toothed, and up to 3 mm across. This species fruits on wood. It is found in Ontario and eastern Canada, but is not common.

Oxyporus populinus

Fruitbodies are bracket-form, broadly attached, up to 10 cm wide, semi-circular to elongate, white to cream, velvety, becoming smooth, and often green from moss or algae. Pores are white to cream, tiny, 4–7 per mm, and in layers (perennial), each up to 4 mm thick. Widespread and not uncommon, this species fruits on living hardwoods, especially maple.

Phlebia radiata

Fruitbodies are ellipsoid to circular or irregular in outline and form flat, sheet-like growths that are white to pinkish or salmon to orange-red, and with raised ridges radiating from the centre to the margin. Flesh is thin, waxy, soft, and gelatinous. Widespread and common, this species fruits on debarked logs and stumps.

Phlebia incarnata

Fruitbodies are flat or shelving, sometimes overlapping, soft to leathery, up to 6 cm across at the base by 5 cm wide, coral-pink, and fading to whitish. Underside is irregular or with net-like folds or shallow, elongate pores. Widespread but not common, this species fruits on hardwood logs or branches. Also known as *Merulius incarnata*.

Phlebia tremellosa

Fruitbodies are flat or narrowly shelving, soft, gelatinous, whitish, with a hairy to woolly upper surface and pinkish undersurface, and with shallow pores or net-like folds. Flesh is thin, white, and waxy to gelatinous. Pores are 1–1.5 mm across. Widespread and common, this species fruits on dead wood. Also known as *Merulius tremellosus*.

Thelephora caryophyllea

Carnation Fungus

Fruitbodies are funnel-shaped, 1.5–5 cm tall, up to 5 cm across, with a central stalk and a radially ridged surface, leathery, pliant, purple-brown, and streaked with dark fibres. Secondary caps or lobes develop inside the primary cap to give the carnation-like appearance on which the species name is based. Widespread and locally common, Carnation Fungus fruits on the ground under pines.

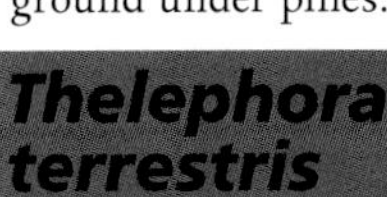

Thelephora terrestris

Earth Fan

Fruitbodies are sessile or with short, stem-like bases, fan-shaped, clustered, often fused, fawn to dark brown, with a whitish margin, covered with shaggy hairs, thin, and splitting along the margin. Underside is more or less smooth and brown. Widespread and common, Earth Fan fruits on the ground in forest nurseries, woods, etc.

Thelephora palmata

Fruitbodies are erect, 2–6 cm tall by up to 3 cm wide, composed of many flattened branches, leathery-soft, purple-brown, with whitish tips at first, drying red-brown, and with a fetid odour. Widespread and not uncommon, this species fruits on moist ground in fields and woods.

Cotylidia diaphana

Fruitbodies are stalked, white to cream, funnel-shaped, thin, tough, pliant, striate, faintly zonate, sometimes split at the margin, 2–4 cm tall, and up to 4 cm across. Stalks are thin, white, downy, and 1–3 mm wide. Widespread and not uncommon, this unusually attractive species fruits on the ground in deciduous woods.

Stereum ostrea

False Turkey Tail

Fruitbodies are shelving and overlapping, up to 8 cm across at the base by 5 cm wide, thin, clothed with radially arranged silky hairs, zonate, and buff-yellow to wood-brown or cinnamon-brown. Underside lacks pores. Widespread and common, False Turkey Tail fruits on dead wood.

Stereum hirsutum

Fruitbodies are shelving, overlapping, up to 2 cm long by 2 cm wide, buff, becoming grey, leathery, thin, covered with stiff hairs, and zonate. Underside is buff, ageing grey. Widespread, this species fruits on stumps of deciduous trees.

Photo: Scott Redhead.

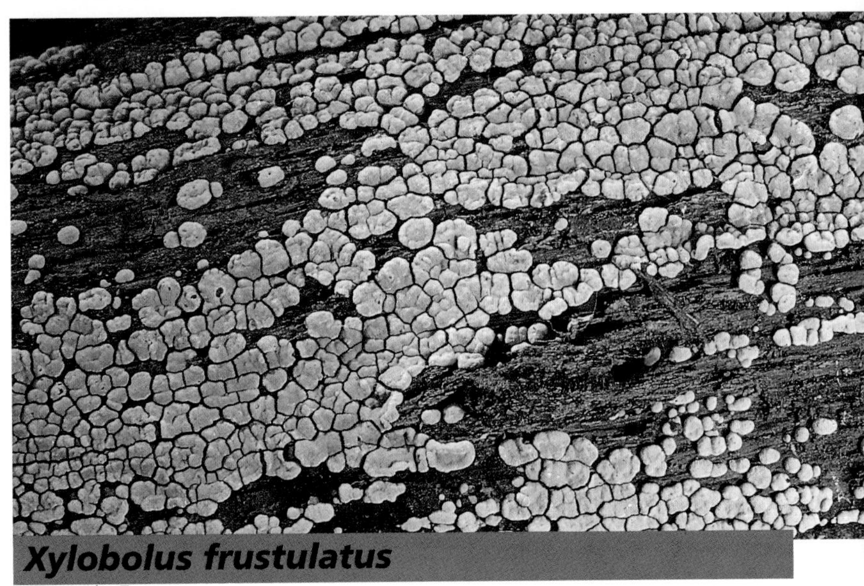

Xylobolus frustulatus

Ceramic Fungus

Fruitbodies form flat crusts, up to 20 cm wide, that run along the surface of logs. Crust is composed of irregular, tile-like segments that are dirty white to pinkish-white or pale yellow-brown, ageing to greyish-brown. Widely distributed but not common, distinctive and unusual, Ceramic Fungus prefers old, debarked oak logs to fruit on.

Peniophora rufa

Fruitbodies are flat, discoid, scattered, tough-fleshy, 1–2 mm thick, 2–4 mm across, vinaceous-brown to red-brown, with a raised margin and coarsely wrinkled surface. Widespread and not uncommon, this species fruits on dead branches of poplars and willows.

Photo: Greg Thorn.

BOLETES
(Sponge Mushrooms)
Basidiomycota—Hymenomycetes

At first glance boletes look like gill mushrooms, because they have well-developed caps and stalks (see Fig. 11, p. 181). Underneath, however, boletes have **tubes** instead of gills, and in this feature they resemble bracket fungi. Boletes differ from bracket fungi in that they are **soft** and **fleshy** and **rot** quickly. They are sometimes called "**sponge fungi**." Bracket fungi, on the other hand, are leathery or woody, persist for long periods, and are often referred to as polypores.

In boletes the spore mother cells (basidia) form a layer (hymenium) that lines the inside of the tubes. Tubes can be very shallow or more than a centimetre deep in some species. The mouths of the tubes appear as pores on the underside of the fruitbody. Pores vary in size and shape from species to species (Fig. 9, p. 157): in some species pores are circular; in others angular or labyrinthiform; in some, the pores are radially arranged; and in others irregular. Pores can be more than a centimetre across or so tiny they are hard to see without a magnifying glass.

As a group, boletes are highly prized as edibles and many are listed as choice. Despite this reputation, they are often overrated and have little flavour. Some, however, are truly delicious and amongst these is *Boletus edulis*, King Bolete (p. 163). While most boletes are edible, some, such as *Boletus satanas*, are poisonous and others, such as Bitter Bolete (*Tylopilus felleus*, p. 168), are so bitter as to be inedible.

When collecting boletes you'll quickly notice that, more often than not, insects find the fruitbodies before you do and the mushrooms are often riddled with larvae and their tracks (Fig. 9). Within moderation, however, this is not a problem for many collectors who happily consume larvae and consider them as an animal protein bonus.

Identification of Boletes

With a little experience a few of the common boletes can be recognized on sight. A number of characteristics are used to make identifications in this group. **Shape**, **size** and **colour** are always important. (See p. 181 for information on cap shapes.) Also, chewing a tiny morsel establishes whether it is **bitter** or **not**. The flesh and/or pores of many boletes **discolour** when bruised, and the shade and speed of **staining** is a useful diagnostic tool. Pore size, shape and depth are important features and whether or not the pores are in rows or radially arranged can help. Is the cap smooth or rough, slimy or dry? **Location** is an additional aid. A number of boletes favour association with specific trees, e.g., *Suillus americanus* (p. 177) is only found with white pine. It also helps to know the colour of the spores *en masse*. This information is obtained from a **spore print**. To make a spore print, cut the cap off a mature bolete and place it on a sheet of white paper. Cover with a cup or bowl and leave overnight. The following day, the discharged spores will leave an imprint on the white paper! The colour will vary according to the thickness and freshness of the print. There are a number of insects called Mycetophilidae (fungus lovers), however, that specialize in stinging fruitbodies and laying their eggs in the flesh. So, when making a spore print from a bolete cap, you might end up the next day with a writhing heap of white larvae with black heads. It is best to inspect the mushrooms for the presence of larvae before using them for spore prints.

Comments on the Genera of Boletes

Boletes have had a checkered history. The various species have moved back and forward amongst different genera and each bolete has been known by several different names over the years. The species names themselves are undergoing considerable revisions. If you pick up a selection of three currently available books on mushrooms, you might find the same fungus under three different scientific names. This example suggests that there is a lot of uncertainty and difference of opinion about the classification system that should be applied to boletes. Hopefully, many of the questions will be answered by the new molecular techniques that are now being applied to the fungi. Until these matters are resolved, however, we will have to bear with the contradictions. On the next page, I have given some brief comments that might aid in distinguishing one bolete genus from another.

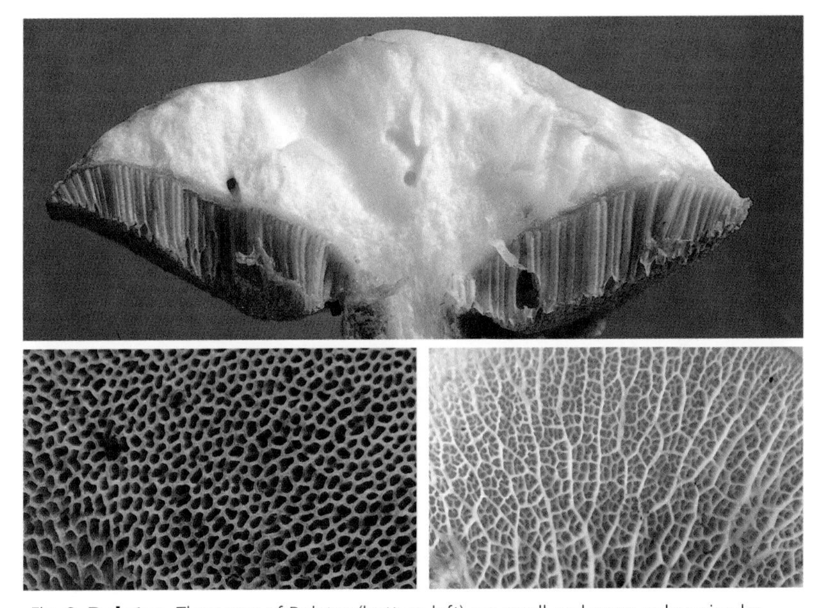

Fig. 9. **Boletes**. The pores of Boletus (bottom left) are small and more or less circular. In Gyrodon merulioides (bottom right), the pores are large, angular and more or less radially arranged. A lengthwise cut through the cap of Suillus (top) shows tubes that are about a centimetre deep. Note the larval tracks. Insects usually find the boletes before you do and most boletes have already been attacked at the time of picking. Cook or refrigerate as soon as possible after collecting. The hymenium (spore mother cells) lines the inside of the tubes. The spores are shot off into the centre of the tube, and then drop by gravity through the pore to be carried off by gentle breezes.

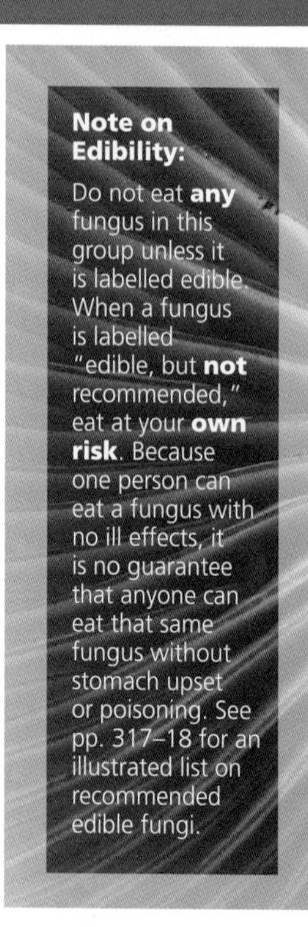

Strobilomyces:

The single species treated here, Old Man of the Woods (*S. strobilaceus*, p. 159), is recognized by its black, shaggy cap and stalk and grey pore surface.

Gyrodon:

The single species, Ash Bolete (*G. merulioides*, p. 177), is recognized by its large, angular, radial pores, its off-centre stalk and its association with ash trees.

Gyroporus:

Blue-Staining Bolete (*G. cyanescens*, p. 165) is the only species treated here. It has a yellow spore print, is off-white and stains brilliant blue on bruising.

Leccinum:

Species in this large genus are recognized by the dark dotted stalks. The dots are mostly black or dark brown, but are sometimes paler. Some species of *Suillus*, e.g., Granular-Dotted Bolete (*S. granulatus*, p. 174), have light-coloured dots on the stalks. Also, the tubes of *Leccinum* are never yellow, as they can be in *Sullius*.

Tylopilus:

The very common Bitter Bolete (*T. felleus*, p. 168) has a pink spore print, and the whitish pores turn pinkish in age. There is a network of ridges in the upper half of the stalk of this species and the flesh has a very bitter taste.

Chalciporus:

The only species, Peppery Bolete (*C. piperatus*, p. 170), is recognized by its small size, peppery taste, cinnamon cap and bright yellow flesh.

Porphyrellus and Austroboletus are represented by a single species each. Both genera have red-brown spore prints. *Porphyrellus* bruises blue and *Austroboletus* doesn't stain. Some mycologists place both species in *Porphyrellus*.

Boletellus is represented by two species with striate spores. Russell's Bolete (*B. russellii*, p. 160) is a very distinctive species, with strongly fluted (lacunose) stalks, that grows under oaks. *B. chrysenteroides* (p. 160) is very similar to *Boletus chrysenteron*, but it has striate spores and lacks the red pigment in the cracks in the cap, which *Boletus chrysenteron* has.

Phylloporus is represented by a single species, Gilled Bolete (*P. rhodoxanthus*, p. 159), which is distinctive in having bright yellow gills rather than pores and is, therefore, more likely to be confused with a brown-spored gill fungus.

Boletus and Suillus:

If the bolete you find is in none of the above genera, then it might well be either a *Boletus* or a *Suillus* but the distinctions here are hazy for beginners. In general, if it has a ring and/or large, angular, radial pores then it is a *Suillus*. If it is slimy, then it is also likely to be a *Suillus*. Most *Boletus* are dry or, at best, slippery when wet and have small pores.

Phylloporus rhodoxanthus

Gilled Bolete

Caps are 2–6 cm across, convex, becoming flat, dry, velvety, and chestnut-brown to red-brown. Flesh is yellow. Stalks are up to 6 cm tall, pale yellow-buff, and with a reddish tinge. Gills are decurrent, bright yellow, well-spaced, bruising blue, and often forked or with transverse connectives. Spore print is olive, drying yellow-brown. Gilled Bolete fruits on the ground in mixed woods. Edible.

Strobilomyces strobilaceus

Old Man of the Woods

Caps are 5–15 cm across, convex, dry, and with thick, fibrous, black scales on a greyish to grey-brown background. Flesh is white, staining red to black. Tubes are white, and stain red, then black. Stalks are up to 13 cm tall by 3 cm wide, coloured as the cap, and shaggy. Spore print is black. Old Man of the Woods fruits on the ground or dead wood. Also known as *S. floccopus*. Edible.

Photo: Brian Shelton.

Boletellus russellii

Russell's Bolete

Caps are 3–9 cm across, convex, dry, felty to scaly, and light brown to ochre, with a pinkish tint, often breaking up into patches. Flesh is yellowish, not staining. Tubes are cream to mustard-yellow, not staining. Pores are large and angular. Stalks are up to 20 cm tall by 2.5 cm wide, red-brown, and with deep longitudinal flutes forming a shaggy, raised network. Spore print is dark brown. Widespread and not uncommon, Russell's Bolete fruits under hardwoods, especially oaks. Edible.

Boletellus chrysenteroides

Caps are 2–10 cm across, convex, dry, hairy, rich brown to olive-brown, and becoming cracked in a mosaic-like pattern over the surface to expose the flesh. Flesh is cream to pale yellow, staining blue. Odour is pleasant. Tastes mild. Tubes are lemon-yellow, with a greenish tinge, staining blue, and ageing brown. Stalks are up to 12 cm tall by 1.5 cm wide, brown, and with darker streaks. Spore print is olive-brown. This species fruits on the ground in mixed woods. Similar to Red-Cracked Bolete (*Boletus chrysenteron*, p. 161) but it lacks red in the cracks of the cap and is distinguished microscopically by striate spores. Edible.

Boletus chrysenteron

Red-Cracked Bolete

Caps are 3–10 cm across, convex to flat, dry, velvety-hairy, and olivaceous-brown to reddish-purple, with mosaic-like, reddish cracks over the surface. Flesh is yellowish, staining bluish slowly. Tubes are large and greenish-yellow, staining blue to brownish. Pores are up to 1 mm wide and angular. Stalks are up to 9 cm tall by 1.5 cm wide, striate, and yellow to red. Spore print is olive, drying to ochre. Red-Cracked Bolete fruits under hardwoods. Also known as *Xerocomus chrysenteron*. Edible.

Boletus affinis var. *maculosus*

Spotted Bolete

Caps are 5–10 cm across, convex, smooth, dry, yellow-brown, and becoming olive-brown, dotted with lighter spots. Flesh is white, not staining. Tubes are white, ageing yellowish to buff, and bruising brownish. Pores are 2–3 per mm and round to angular. Stalks are up to 8 cm tall by 2.5 cm wide. Spore print is bright yellow-brown. This species fruits under hardwoods. Also known as *Xanthoconium affine*. Edible.

Boletus ornatipes

Caps are 5–15 cm across, convex, velvety, sticky when wet, and olive-brown, becoming grey-olive to smoky. Flesh is bright yellow, mildly bitter, and not staining. Tubes are up to 8 mm deep, lemon-yellow, and staining orange-yellow. Pores are 2 per mm. Stalks are up to 15 cm tall, bright yellow to dull yellow, and with a prominent surface network. Spore print is olive-brown, drying yellow-brown. This species fruits under hardwoods. Edible.

Boletus parasiticus

Parasitic Bolete

Caps are 3–8 cm across, convex to broadly convex, dry, minutely hairy, becoming smooth, and olive-yellow to ochre. Flesh is white to pale yellow. Tubes are golden-brown to olive. Stalks are up to 10 cm tall by 1.5 cm wide. Spore print is olive-brown. This species is parasitic on the fruitbodies of *Scleroderma* (p. 92).

Boletus pallidus

Pale Bolete

Caps are 5–15 cm across, convex, dry, smooth, and dirty white to grey-brown, ageing buff. Flesh is whitish, sometimes staining yellowish, and mild or slightly bitter-tasting. Tubes are white, ageing olive, and staining blue. Pores are medium to large and angular to irregular. Stalks are up to 15 cm tall by 2 cm wide and off-white. Spore print is olive-brown. Pale Bolete fruits under hardwoods, especially oaks. Edible.

Boletus bicolor

Two-Coloured Bolete

Caps are 5–25 cm across, convex, smooth, dry, red to purple-red, and mottled with yellow. Flesh is yellow, staining blue, and then fading. Tubes are up to 15 mm deep, yellow, ageing olive-brown, and bruising blue. Pores are angular and 1 mm across. Stalks are up to 15 cm tall by 6 cm wide, yellow near the cap, red below, and staining blue. Spore print is olive-brown. Two-Coloured Bolete fruits under hardwoods. Edible.

Boletus edulis

King Bolete

Caps are 10–20 cm across, convex, slippery when wet, smooth, and ochre to red-brown. Flesh is white, not staining. Tubes are white, becoming yellowish, tinged with green, and staining yellowish. Pores are tiny and round. Stalks are up to 15 cm tall by 3 cm wide, with a bulbous base, white to ochre, and with net-like markings, especially near the apex. Spore print is olive-brown. King Bolete fruits on the ground under conifers or in mixed woods. Edible.

Boletus subvelutipes

Caps are 4–18 cm across, convex, dry, velvety to smooth, and yellow-brown to brick-red or red-brown. Flesh is yellow, rapidly staining blue-black, and mild-tasting. Tubes are yellow to brown, rapidly staining blue-black. Pores are orange-red, small, and round. Stalks are up to 12 cm tall by 5 cm wide, smooth to powdery, and coloured as the cap or paler. Spore print is olive-brown. This species fruits on the ground in mixed woods. Edible, but **not** recommended as some red-pored boletes are **poisonous**.

Boletus badius

Bay-Brown Bolete

Caps are 4–11 cm across, convex, slippery when wet, smooth to minutely hairy, and bay-brown to red-brown or golden-brown. Flesh is white to yellowish, staining blue, and then fading. Tubes are yellowish to greenish-yellow and staining greenish-blue. Pores are large, up to 1.5 mm, and round to angular. Stalks are up to 10 cm tall by 2.5 cm wide, often curved, and coloured as the cap. Spore print is olive-brown to ochre-brown. Bay-Brown Bolete fruits on the ground or on rotting stumps in conifer woods. Also known as *Xerocomus badius*. Edible.

Gyroporus cyanescens

Blue-Staining Bolete

Caps are 3–11 cm across, convex, dry, hairy to woolly, and dirty-white to tan. Flesh is white and staining rapidly to brilliant blue. Tubes are white to yellowish, staining indigo. Pores are small and round. Stalks are up to 10 cm tall by 3.5 cm wide, hairy, and coloured and staining as the cap. Spore print is yellow. This species fruits under hardwoods, often beside woodland trails. Edible.

Boletus griseus

Grey Bolete

Caps are 4–12 cm across, convex, dry, dull, rough, and grey to grey-brown. Flesh is pale grey to yellowish-grey, staining purplish. Tubes are grey to grey-brown. Pores are 1–2 per mm and not staining. Stalks are up to 10 cm long by 2.5 cm wide, with a surface network, and yellowish towards the base. Spore print is olive-brown. Widespread and not uncommon, Grey Bolete fruits under hardwoods, particularly oak and beech. Edible.

Leccinum holopus

Caps are 3–10 cm across, convex, dry to sticky, smooth, and off-white, with a greenish tinge, to pale smoky-grey. Flesh is white, not staining. Stalks are up to 12 cm tall by 2 cm wide, pallid, and dotted with dark brownish tufts. Tubes are up to 25 mm deep, pallid, and bruising brownish. Pores are whitish, becoming brownish. Spore print is cinnamon-brown. This species fruits on the ground in wet spots in woods. Edible.

Leccinum scabrum

Birch Bolete

Caps are 5–20 cm across, convex, slippery when wet, smooth to velvety, and tan to smoky or blackish-brown. Flesh is whitish, not staining. Tubes are white to greyish, staining darker. Pores are tiny and round. Stalks are up to 15 cm tall by 2 cm wide, white to grey or smoky, and with raised, grey to grey-black dots. Spore print is brown to olivaceous-brown. Birch Bolete fruits on the ground mostly under birch. Edible.

Leccinum aurantiacum

Orange Bolete

Caps are 5–20 cm across, convex, dry, minutely hairy or scaly, and orange-yellow to rusty or orange-brown. Flesh is white to brownish and staining wine-red, then grey to black. Tubes are white to grey. Pores are tiny and round. Stalks are up to 20 cm tall by 5 cm wide, white to brownish, and with red-brown or black dots. Spore print is light brown. Orange Bolete fruits under birch and poplar. Edible.

Leccinum atrostipitatum

Dark-Stalked Bolete

Caps are 7–20 cm across, convex, dry, minutely hairy to fibril-streaked, and buff to yellow-orange or ochre. Flesh is white, staining grey, then black. Tubes are grey-brown, staining as flesh. Pores are small, round, and smoky to blackish-brown. Stalks are up to 20 cm tall by 3.5 cm wide, whitish, and with many raised, black dots. Spore print is yellow-brown to olive. Dark-Stalked Bolete fruits on the ground in mixed woods. Edible.

Leccinum insigne

In field characteristics, *L. insigne* is very similar to Orange Bolete (*L. aurantiacum*, p. 166), but it differs in that the flesh does not bruise reddish. *L. insigne* is found under poplars and pines and is a common bolete in northern arboreal forests. Edible.

Boletus subglabripes

Caps are 2–12 cm across, convex, dry, smooth, and ochre to cinnamon or red-brown. Flesh is yellowish, not staining. Tubes are yellow to yellow-green to olive. Pores are small and round to angular. Stalks are up to 10 cm tall by 2 cm wide and tinged with red or purple. Spore print is olive-brown. This species fruits on the ground in mixed woods. Also known as *Leccinum subglabripes*. Edible.

Photo: Greg Thorn.

Tylopilus eximius

Caps are 8–25 cm across, convex, dry, powdery to smooth, and purple-brown to chocolate-brown. Flesh is greyish to pale purple-grey, not staining. Tubes are brown. Pores are tiny and round. Stalks are up to 12 cm tall by 2.5 cm wide and purplish-grey to purplish-brown. Spore print is buff with a purplish tinge to dark brown. This species fruits under conifers. Edible.

Tylopilus felleus

Bitter Bolete

Caps are 8–25 cm across, convex, dry, smooth to powdery, and tan or yellow-brown to red-brown. Flesh is white and not staining, and tastes very bitter. Tubes are white to tan, staining pinkish to brown. Pores are small and angular. Stalks are up to 15 cm tall by 3 cm wide, broader at the base, and with the surface network near the cap. Spore print is pink. Bitter Bolete fruits on the ground in mixed or coniferous woods.

Photo: Brian Shelton.

Austroboletus gracilis

Caps are 4–8 cm across, convex to flat, dry, smooth to slightly granular, and tan to reddish-brown. Flesh is white, not staining. Tubes are whitish to pinkish-buff. Pores are small and round. Stalks are up to 12 cm tall by 1 cm wide, coloured as the cap or paler, and powdery or striate. Spore print is rich brown to vinaceous-brown. This species fruits under conifers, particularly hemlock. Also known as *Porphyrellus gracilis*. Edible.

Tylopilus chromapes

Caps are 4–12 cm across, convex to flat, dry, hairy to smooth, and pinkish-red to tan. Flesh is white, pink below the skin, and not staining. Tubes are white to flesh-coloured, ageing brown. Pores are tiny and round. Stalks are up to 10 cm tall by 2 cm wide, white with pinkish tints, and with a bright yellow base, covered with raised, reddish dots. Spore print is pinkish-brown. This species fruits on the ground in woods. Also known as *Leccinum chromapes*. Edible.

Chalciporus piperatus

Peppery Bolete

Caps are 2–8 cm across, convex to flat, slippery when wet, smooth to hairy near the margin, and yellowish-brown to cinnamon or red-brown. Flesh is white to yellowish, bruising greyish to brownish, and with a peppery taste. Tubes are reddish-brown. Pores are large and angular. Stalks are slender, up to 8 cm tall by 8 mm wide, coloured as the cap, striate, and yellow at the base and inside. Spore print is cinnamon. Peppery Bolete fruits under conifers. Also known as *Boletus piperatus*.

Suillus luteus

Slippery Jack

Caps are 4–12 cm across, convex, slimy, smooth, and yellow-brown to red-brown. Flesh is yellow, not staining. Tubes are yellow. Pores are small and angular. Stalks are up to 10 cm tall by 2.5 cm wide and dotted in the upper part. Ring is persistent, prominent, and flaring. Spore print is dull ochre. Slippery Jack fruits under conifers. Edible, but causes **stomach upset** in some people.

Suillus salmonicolor

Slippery Jill

Caps are 4–10 cm across, convex to broadly convex, with a slimy surface, and dingy yellowish to red-brown. Flesh is whitish or with a yellow tinge. Tubes are yellowish to ochre. Pores are less than 1 mm. Stalks are up to 7 cm tall by 1 cm wide, cylindrical, and uniformly covered with dark brown dots above and below the ring. Ring is narrow and collapsing, sometimes disappearing. Slippery Jill fruits under conifers, especially white pine. Also known as *S. subluteus*. Edible.

Suillus umbonatus

Caps are 3.5–6 cm across, broadly convex, with a low knob, smooth, slimy, and brown to olive-buff. Flesh is pale buff, staining pinkish. Tubes are light yellow and bruise brown. Pores are angular, radially arranged, and up to 2 mm tall and 2 mm wide. Stalks are up to 4 cm tall by 7 mm wide and covered with scattered red dots. Ring is brown to pinkish-brown and persistent. Spore print is whitish to pale cinnamon. This species fruits under jack pine in our region.

Suillus cavipes

Hollow-Stemmed Bolete

Caps are 4–12 cm across, convex, with a low knob, dry, tan to rust-brown, and covered with reddish hairs or scales. Flesh is yellow, not staining, and tastes mealy to bitter. Pores are large, angular, dingy-yellow to ochre, radially arranged, and not staining. Stalks are up to 8 cm tall by 1 cm wide, with a swollen base, hollow, and coloured as the cap. Ring is white to ochre and disappearing. Spore print is olive-brown. Fruits under larch. Also known as *Boletinus cavipes*. Edible.

Suillus sinuspaulianus

Caps are 3–13 cm across, convex, becoming flat, smooth but fibre-streaked, sticky, and rich red-brown, fading to orange-brown. Flesh is yellow to buff, not staining. Pores are large, angular, and radially arranged. Stalks are up to 12 cm tall by 3 cm wide, yellow-brown to greyish-brown, and spotted with red. Ring has a raised network above it. Spore print is chocolate-brown to purple-brown. This species fruits under conifers. Also known as *Fuscoboletinus sinuspaulianus*. Edible.

Photo: Greg Thorn.

Suillus paluster

Larch Bolete

Caps are 2–7 cm across, convex to flat, with a low knob, dry, radially fibre-streaked to floccose, and wine-red. Flesh is yellow, not staining. Tubes are decurrent and pale yellow to ochre. Pores are large and angular. Stalks are up to 5 cm tall by 6 mm wide, powdery, and coloured as the cap. Spore print is purple-brown. This species fruits under larch. Also known as *Fuscoboletinus paluster*. Edible.

Photo: Greg Thorn.

Suillus spraguei

Painted Bolete

Caps are 5–10 cm across, convex to broadly convex, dry, and covered with reddish scales on a yellow base. Flesh is yellowish, staining red to brownish. Tubes are yellowish, staining brownish. Pores are large, angular, and radially arranged. Stalks are up to 8 cm tall by 1.5 cm wide, coloured as the cap, and scaly. Ring is persistent. Spore print is ochre. Painted Bolete fruits under conifers, especially white pine. Formerly known as *S. pictus*. Edible.

Suillus laricinus

Caps are 3–12 cm across, convex to flat, slimy, smoky-brown to blackish-brown, and bleaching to whitish. Flesh is whitish, bruising blue. Tubes are up to 9 mm deep, whitish, and ageing smoky to brownish. Pores are large, up to 3 mm across, angular, and radially arranged. Stalks are up to 6 cm tall, with a ring, and smoky to brownish. Spore print is purple-brown. This species fruits under larch. Also known as *Boletus aeruginascens* and *S. viscidus*.

Suillus granulatus

Granular-Dotted Bolete

Caps are 2–13 cm across, convex to broadly convex, smooth, slimy, tan to cinnamon, and ageing white with brown stains. Flesh is yellowish, not staining. Tubes are white to dull yellow, not staining. Pores are up to 1 mm across and with an irregular shape. Stalks are up to 9 cm tall by 2.5 cm wide, white to yellowish, and dotted tan-brown in the upper half. Spore print is yellow-brown. Granular-Dotted Bolete fruits under conifers (mainly pines or spruce). Edible.

Suillus brevipes

Short-Stalked Bolete

Caps are 5–10 cm across, convex, smooth, slimy, and buff to cinnamon or rust-brown. Flesh is white, becoming yellow, and not staining. Tubes are up to 10 mm deep and dingy yellow, ageing brownish. Pores are 1–2 per mm. Stalks are up to 5 cm tall by 2 cm wide, white, and becoming yellowish, especially near the apex. Spore print is ochre-brown. Short-Stalked Bolete fruits under pines. Edible.

Porphyrellus porphyrosporus

Caps are 5–15 cm across, convex, olive-brown to almost black, dry, and velvety, becoming smooth. Flesh is white, becoming red-brown, staining blue-green, and tastes mild. Tubes are greyish-white and ageing or staining dark brown. Pores are small. Stalks are up to 10 cm tall by 3 cm wide, coloured as the cap, staining brown, and with a swollen, whitish base. Fruits on the ground in mixed woods. Also known as *Tylopilus pseudoscaber* and *P. pseudoscaber*. Edible.

Suillus placidus

Caps are 3–10 cm across, convex, becoming flat, smooth, slimy when wet, shiny when dry, ivory-white or with a yellowish tinge, and ageing brown. Flesh is white, yellowish near the tubes, and not staining. Tubes are yellowish, not staining. Pores are yellow and 1–2 per mm. Stalks are up to 12 cm tall by 2 cm wide and whitish, with wine-brown blotches. Spore print is dull cinnamon. This species fruits under white pine. Edible.

Suillus acidus

Caps are 4–7 cm across, convex, smooth, slimy, pallid yellow, and bruising brown. Flesh is pallid, not staining, and with an acid taste. Tubes are pale yellow. Pores are pale yellow, small, and not staining. Stalks are up to 8 cm tall by 1 cm wide, with a dotted surface, sticky, and whitish to buff. Spore print is ochre-brown. This species usually fruits under red pine. Edible.

Photo: Greg Thorn.

Suillus americanus

White Pine Bolete

Caps are 3–10 cm across, convex to flat, slimy, and striate or spotted, with flattened, red-brown scales, especially near the margin. Flesh is yellow, staining reddish to brown. Tubes are tan to yellow and radially arranged. Pores are large and angular. Stalks are up to 7 cm tall by 8 mm wide, with brownish dots. Ring disappears. Spore print is ochre-brown. White Pine Bolete fruits under white pine. Edible.

Suillus tomentosus

Caps are 5–10 cm across, convex to flat, slippery when wet, cottony/scaly, becoming smooth in age, and yellow to orange-yellow. Flesh is pallid to yellow, staining blue slowly. Tubes are yellowish to ochre, staining as flesh. Pores are small and angular. Stalks are up to 10 cm tall by 3 cm wide, coloured as the cap, and with red-brown to blackish dots. Spore print is olive. This species fruits on the ground in jack pine forests. Edible.

Gyrodon merulioides

Ash Bolete

Caps are 7–20 cm across, kidney-shaped to irregularly lobed, becoming flat, with an inrolled margin, dry, and yellow-brown to olive. Flesh is tough, yellow, and staining blue-green to brown. Tubes are shallow and yellow to ochre. Pores are radially arranged, large, and variable. Stalks are short, up to 4 cm tall, stout, and off-centre. Ash Bolete fruits under ash. Also known as *Boletinellus merulioides*. Edible.

GILL FUNGI

Scientific nomenclature is fairly stable for some areas of study. For example, in books on birds, you will find that the names in the books you buy today are almost the same as those used by Audubon so many years ago. Unfortunately, the same is not true for fungi, especially gill fungi. You will find that the scientific names used in this book are a bit different from other books you have seen and a lot different from books that were published no more than a decade or two ago. Mycology is a relatively young science and the classification system is still in a state of flux. Recent and more detailed studies have shown that species once considered closely related from their superficial morphology are, in fact, quite different in more fundamental characteristics, including microscopic features. The result of these studies is that a wealth of new scientific names has been added to the literature. Often, before the dust has settled on one study, another study has re-evaluated the situation and we then have changes to the changes. Indeed, sometimes we seem to be approaching chaos. We now have several scientific names for each mushroom and it is not always easy to decide which is the "best" name—which is especially frustrating for beginners. The most recent name isn't necessarily the best name because, at the finer level of distinctions between genera, it often boils down to a matter of opinion. So, for the gill fungi in this book, I have often given an alternate name at the end of the description. Feel free to use the one you like best or the one you can remember! Until taxonomists get it firmed up, we shouldn't be too uncompromising about the "correct" scientific name for a species.

In this book the gill fungi are separated into four groups on the basis of the colour of the spore prints: Pink-Spored, Dark-Spored, Brown-Spored and Light-Spored. Spore prints, to determine colour, are obtained as outlined in the section on boletes (see p. 156). The range in colours, and shades of colour, for each group is given in more detail in the introductions to each of the four groups. The light-spored group (pp. 231–315) is by far the largest. To make it a little more manageable, I have written short introductions for the larger, more common or more interesting genera or groups of genera. In these introductory paragraphs, I draw attention to identification features, edibility, related fungi and other points of interest or importance.

Once we have established (or guessed at) the spore colour, there are a number of features associated with the fruitbodies that are used as aids in identification. These features—the gill attachment, cap shapes and structure of a mushroom—are summarized in Fig. 11 (p. 181). The size and colour of the cap and whether it is smooth or scaly, dry or slimy is also helpful in identifying mushrooms. The spacing of the gills (Fig. 10, p. 180) is useful, too. It should be remembered, however, that gill spacing can change with age and the spacing in a young cap might appear quite different from that of a mature, expanded cap. Also, as a first step in identification, it is important to note in mature fruitbodies whether the gills are free or attached to the stalk.

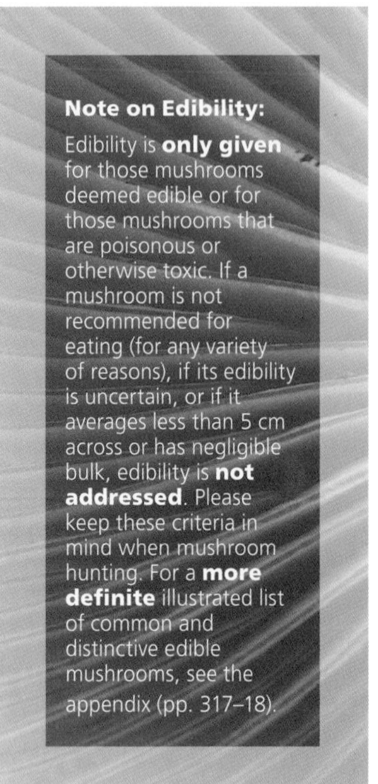

Note on Edibility:

Edibility is **only given** for those mushrooms deemed edible or for those mushrooms that are poisonous or otherwise toxic. If a mushroom is not recommended for eating (for any variety of reasons), if its edibility is uncertain, or if it averages less than 5 cm across or has negligible bulk, edibility is **not addressed**. Please keep these criteria in mind when mushroom hunting. For a **more definite** illustrated list of common and distinctive edible mushrooms, see the appendix (pp. 317–18).

In a few mushrooms, such as *Amanita* (pp. 231–41), the young fruitbody is contained within a membranous envelope called the "**universal veil**." When the growing mushroom expands and breaks out, the veil remains as a **cup** (volva) encasing the base of the stalk. Sometimes parts of the veil are carried up on the cap as **remnants**, and these remnants give a patchy or warty appearance. In some mushrooms, a second veil, called the "**inner veil**," stretches just below the gills from the edge of the cap to the stalk. When this veil breaks, it appears as a **ring** (annulus) around the stalk. Sometimes parts of the inner veil remain attached to the margin of the cap (**appendiculate**). All of these features are useful in identifying mushrooms.

Chemical tests are also used to aid in the identification of mushrooms. In this field guide, I do not suggest these tests for use. One simple chemical test, however, has proved particularly useful, from time to time, to confirm the identity of some light-spored mushrooms. If you have ever taken a biology course, you will probably remember the "starch" test. In iodine, starchy compounds turn black; spores of some fungi react positively to the iodine test. Mycologists use iodine in "Melzer's" reagent (0.5 g iodine, 1.5 g potassium iodide, 20 g chloral hydrate, 20 ml water). If the spores turn blue-grey to black, then they are referred to as "amyloid." In general, the spores are examined microscopically for this change, but a large mass of spores can be tested on a glass slide by visual inspection against a white background.

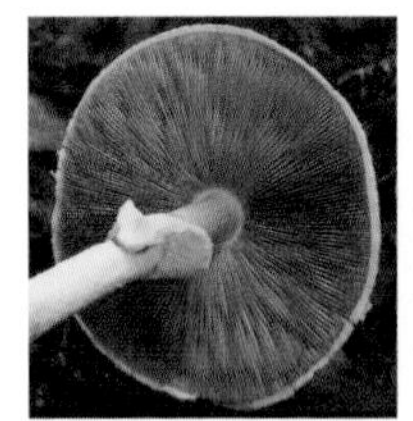

Fig. 10. **Gill Spacing**. Features associated with gills, such as colour, depth, spacing, etc., are valuable in identifying mushrooms. In *Agaricus sylvicola* (far left), the gills are crowded together. In *Mycena leaiana* (second from left), the gills are close but not crowded. In *Mycena pura* (second from right), the gills are well-spaced. In the two *Mycena* spp., you can easily see the secondary and tertiary development of gills, i.e., smaller gills originating at the edge and running between the primary gills. In *Marasmiellus candidus* (far right), the gills are far apart (distant). Broader gills and tighter packing increases the spore-producing surface dramatically. *Marasmiellus* has shallow gills and they are far apart. It is a very successful colonizer of twigs in West Coast rainforests. So! It is clear that there are factors other than spore production involved in the success of a species.

Fig.11. FEATURES OF MUSHROOMS USED FOR IDENTIFICATION

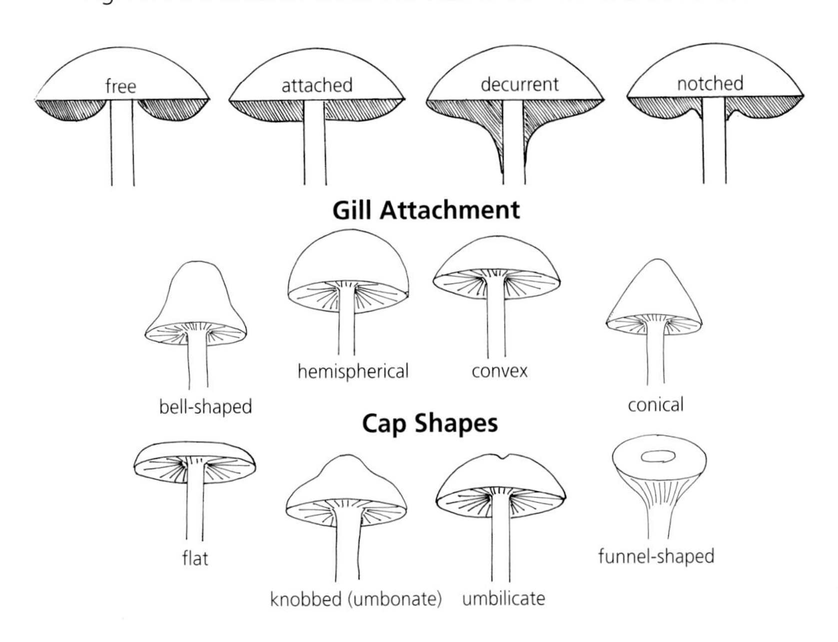

STRUCTURE OF A MUSHROOM

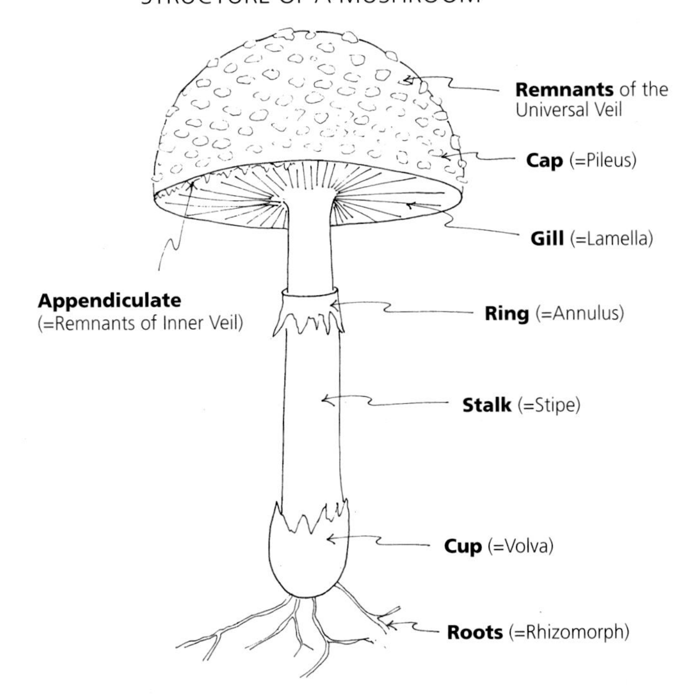

MUSHROOMS WITH PINKISH SPORE PRINTS

Note on Edibility:

Do not eat **any** mushroom in this group unless it is labelled edible. See p. 180 for further information on edibility.

The spore colours of this group are broader than the name indicates and can be pink, salmon, pinkish-buff, flesh-coloured or pinkish-brown. Sometimes spore colour will be indicated by a pinkish tinge to the gills in mature specimens or as a pinkish deposit on the caps of adjacent mushrooms in the group.

There are very few genera of pink-spored fungi, and they are relatively easy to tell apart. *Nolanea* spp. (pp. 183–84) and *Leptonia* spp. (p. 185) are segregates from the genus *Entoloma* (p. 183 and p. 189) and are still included in that genus by some authors. Microscopically, all three genera have distinctive angular spores. *Nolanea* contains species with a conical cap that are often brightly coloured, with the cap prolonged into a narrow, elongated apex. These species are sometimes call "unicorn" fungi. *Leptonia* spp. are also small, but they are broadly convex to flat and often depressed or umbilicate at the centre. The species formerly known as *Clitopilus abortivus* has recently been moved into the *Entoloma* genus (p. 189), because it has angular spores whereas *Clitopilus prunulus* has cylindrical spores. The genus *Rhodocollybia* (p. 190), a segregate of the *Collybia* complex in the light-spored category, gives pinkish-buff spore prints.

The *Entoloma* genus is noteworthy because the larger species, such as *E. lividum* and *E. clypeatum* (p. 183), are thought to have caused a number of **poisonings** of people who had mistaken them for edible species.

Key to Genera of Pink-Spored Mushrooms

The spore colour of this group is broader than the name indicates and can be pink, salmon, pinkish-buff or pinkish-brown (as mentioned in the introduction). This group is small, and it is not difficult to key to genus once the spore colour is established. Sometimes spore colour will be indicated by a pinkish tinge to the gills in mature specimens or as a pinkish deposit on the caps of adjacent mushrooms in the group.

FB = fruitbodies

1. Gills reduced to ridges or lacking, FB grey-black .... ***Craterellus***
1. Gills well developed **2**

2. FB shelving, stalks absent, on wood .... ***Phyllotopsis***
2. FB with cap and central stalk **3**

3. Gills free (not attached to stalk) **4**
3. Gills attached to stalk **5**

4. Cups present at base of stalk .... ***Volvaria***
4. Cups lacking .... ***Pluteus***

5. Gills running down stalk (decurrent) .... ***Entoloma, Clitopilus***
5. Gills not decurrent **6**

6. Caps small (average < 3 cm in diameter) **7**
6. Caps larger (average > 4 cm in diameter) **8**

7. Caps conical to bell-shaped .... ***Nolanea***
7. Caps convex to flat, often with a central depression .... ***Alboleptonia, Leptonia***

8. Caps pinkish-red, with white network over cap .... ***Rhodotus***
8. Caps not pinkish-red **9**

9. Spores salmon-pink, cap with satin sheen .... ***Entoloma***
9. Spores pinkish-buff .... see ***Rhodophyllus, Lepista***

Entoloma clypeatum

Caps are 3–10 cm across, convex, becoming flat, with a broad knob and an irregularly undulating surface, and grey-brown to olive when wet, drying with a silky sheen. Margin is inrolled or turned down. Gills are attached, not crowded, broad, and whitish, becoming pinkish-brown at maturity. Stalks are up to 10 cm tall by 1–2 cm wide and white to greyish-brown. Spore print is pinkish. Widespread and fairly common, this species fruits on the ground in woods. **Poisonous.**

Nolanea lutea

Yellow-Green Nolanea

Caps are 1–2.5 cm across, conical to bell-shaped, greenish-yellow to olivaceous, darker at the apex, smooth to slightly roughened, and weakly striate. Gills are attached, well-spaced, and yellowish-white, becoming pink. Stalks are up to 8 cm tall by 6 mm wide, coloured as the cap, striate, and sometimes twisted. Spore print is pink. Widespread but not common, Yellow-Green Nolanea fruits on the ground in mixed woods. Also known as *Entoloma lutea*.

Nolanea quadrata

Salmon-Coloured Nolanea

Caps are 2–4 cm across, conical to bell-shaped, with a narrow, peaked knob, salmon to orange-salmon, dry, smooth, and faintly striate. Gills are attached, well-spaced, and salmon to salmon-pink. Stalks are up to 8 cm tall by 4 mm wide, smooth, silky, and coloured as the cap. Spore print is pink. Widespread and common, Salmon-Coloured Nolanea fruits on the ground in woods. It is also known as *Entoloma salmoneum* or *E. quadrata*. **Poisonous.**

Nolanea murraii

Yellow Nolanea

Caps are 1–3 cm across, conical to bell-shaped, often with a narrow, elongate knob at the apex, and lemon-yellow. Gills are attached, well-spaced, and pale yellow, becoming flesh-coloured with age. Stalks are up to 8 cm tall, slender, smooth, and coloured as the cap. Spore print is pink. Widespread and not uncommon, Yellow Nolanea fruits on the ground in woods, sometimes on rotten wood. Also known as *Entoloma murraii*.

Leptonia incana

Green Leptonia

Caps are 2–3 cm across, convex to flat, umbilicate, dry, smooth, pea-green to turquoise-green, becoming bright yellow and often radially striate at maturity, and with the odour of mice. Gills are attached, well-spaced, and greenish to yellowish, becoming pale orange-brown with age. Stalks are up to 5 cm tall by 4 mm wide, smooth, shining, and green to yellow-green. Spore print is pink. Readily recognized by its colour, Green Leptonia is common and widespread and fruits on the ground in mixed woods.

Alboleptonia sericella

Caps are 1–4 cm across, convex, becoming flat or depressed, dry, white, sometimes with a pinkish tinge, and smooth to minutely scaly. Gills are attached, distant, and white, ageing pink. Stalks are up to 5 cm tall and slender. Spore print is pink to flesh-coloured and not amyloid. Widespread and not uncommon, this species fruits in wet, grassy spots alongside trails in woods. Also known as *Entoloma sericella*. It has the look of a *Hygrocybe* sp. (pp. 268–77), but its identity is confirmed by the pink spores.

Leptonia formosa

Caps are 1–7 cm across, convex to broadly convex, umbilicate, orange-brown, becoming yellow-brown to dark brown, striate, and rough to minutely scaly. Gills are attached and white to pale yellowish at first, becoming pink. Stalks are up to 8.5 cm tall by 6 mm wide, coloured as the cap, and smooth. Spore print is pink. Widespread and not uncommon, this species fruits on the ground in woods.

Pluteus atricapillus

Deer Mushroom

Caps are 5–12 cm across, bell-shaped to convex, becoming flat, sometimes broadly knobbed, moist to dry, smooth or fibrous-streaked, dark brown to grey-brown or pale tan, fading in age, and paler near the margin. Gills are free, close, broad, and white, becoming flesh-coloured. Stalks are up to 15 cm tall by 1.5 cm wide, smooth, off-white to brown, and fibril-streaked. Spore print is pink. Widespread and very common, Deer Mushroom fruits on rotten wood. Also known as *P. cervinus*. Edible.

Pluteus umbrosus

Caps are 4–9 cm across, convex, becoming flat, with a broad knob, dark brown, radially streaked, and with tiny, blackish scales. Gills are free, close, and white, becoming pinkish. Stalks are up to 9 cm tall by 1.5 cm wide, slender, and coloured as the cap. Spore print is pink. Widespread and not uncommon, this species is recognized by its distinctive markings on the cap. It fruits on well-rotted hardwood logs. Edible.

Pluteus longistriatus

Caps are 2–5 cm across, convex to flat, thin, ash-brown to pale brown, minutely scaly, and radially striate in age. Gills are free, close, broad, and white, becoming flesh-coloured. Stalks are up to 7 cm tall by 1 cm wide and fibre-streaked. Spore print is pale flesh-pink. Widespread but not uncommon, this species fruits on hardwood logs and stumps. The radial striations of the cap make this species distinctive. Edible.

Photo: Greg Thorn.

Pluteus chrysophlebius

Caps are 1–2.5 cm across, hemispherical to flat, becoming depressed near the centre, thin, orange-yellow to yellow-brown, and with darker veins radiating from the centre. Gills are free and white, ageing pinkish. Stalks are up to 4.5 cm tall, slender, yellowish, and white near the base. Spore print is pink. Rarely reported, this species is close to *P. admirabilis*, and fruits on well-rotted hardwood logs, especially hickory.

Photo: Greg Thorn.

Pluteus aurantiorugosus

Caps are 2–6 cm across, convex to flat, brilliant orange-red, paler and striate towards the margin, and fading to orange in age. Gills are white, becoming pink. Stalks are up to 8 cm tall by 8 mm wide. Spore print is pink. This striking mushroom is widespread but rare, and it fruits on rotting hardwood logs.

Photo: Greg Thorn.

Pluteus flavofuligineus

Caps are 2–5 cm across, conical at first, becoming flat, knobbed, yellow-brown, and velvety. Gills are free and white, becoming pinkish. Stalks are up to 7 cm tall, slender, white, smooth, and appearing twisted. Widespread but not common, this species fruits on well-rotted wood and is recognized amongst *Pluteus* spp. by its colour and velvety texture.

Pluteus tomentosulus

Caps are 2.5–7.5 cm across, conical to bell-shaped, becoming flat, knobbed, whitish, tinged pinkish, and cottony to woolly. Gills are free and white, becoming pink. Stalks are up to 10 cm tall by 8 mm wide and white, with a fuzzy surface. Spore print is pink. Widespread but not common, this species is recognized amongst *Pluteus* spp. by its white, cottony cap. It fruits on conifer wood, often in swampy areas.

Pluteus admirabilis

Caps are 1–3.5 cm across, convex, becoming flat, knobbed to depressed, thin, smooth, dry, wrinkled towards the centre, yellow to olive-yellow, and with a striate margin. Gills are free, close, fragile, and whitish to yellowish, then rosy. Stalks are up to 5 cm tall by 4 mm wide, smooth, and yellow, with a white, cottony base. Spore print is pink. Widespread and not uncommon, this species fruits on well-rotted wood.

Clitopilus prunulus

The Miller

Caps are 2–9 cm across, convex to flat or depressed, with a lobed margin, silky, cream to buff or greyish, and with the odour of ground meal. Gills are decurrent, close, narrow, and white, ageing pinkish. Stalks are up to 5 cm tall by 1.5 cm wide, white, and with a cottony surface. Spore print is pink. Widespread and fairly common, The Miller fruits on the ground in woods. The variety *orcellus*, shown here, is similar to the variety *prunulus*, but has a slimy cap, is more robust, and is bone-white. Edible.

Entoloma abortivum

Aborted Entoloma

Caps are 5–10 cm across, convex, becoming flat, dove-grey to grey-brown, dry, smooth, with a wavy margin, and with the odour and taste of ground meal. Gills are decurrent, close, narrow, and grey, becoming pink. Stalks are up to 8 cm tall by 13 mm wide, minutely hairy, and coloured as the cap but paler. Spore print is salmon-pink. Widespread and common, Aborted Entoloma fruits on or near rotting wood. Also known as *Clitopilus abortivus*. Edible.

Rhodotus palmatus

Caps are 2–7 cm across, convex, becoming flat, and flesh-coloured to brick-red, with a raised, white network over the surface. Gills are attached, close, and pink. Stalks are up to 3 cm tall by 6 mm wide, lateral, curved, hairy, and coloured as the cap. Spore print is pink. This beautiful and distinctive mushroom is widespread but rare, and it fruits on dead hardwoods.

Photo: Greg Thorn.

Phyllotopsis nidulans

Caps are 1–8 cm across, kidney-shaped or fan-shaped, bright orange-yellow, fading to buff, dry, shelving, covered with coarse hairs, and with an inrolled margin. Odour is unpleasant. Gills are attached, close, narrow, and yellow-orange. Stalks are absent. Spore print is pinkish-tan. Widespread and not uncommon, this attractive and distinctive orange species fruits on decaying logs.

Craterellus fallax

Horn of Plenty

C. fallax has a pink spore print. The spore print is the only way to distinguish it in the field from *C. cornucopioides* (see p. 248), which has a white spore print. Both are edible and, although thin-fleshed, are considered of excellent quality. Widespread and fairly common, Horn of Plenty fruits on the ground in woods. Edible.

MUSHROOMS WITH DARK SPORE PRINTS

The spore prints of this group are dark brown, purple-brown, purple-black, blackish-brown or black. The intensity of the shade is influenced by the thickness of the spore deposit and whether it is fresh or dry. If a ring is present, the spores often coat the upper surface as a dark band. The common field mushrooms (*Agaricus* spp.) and the inky caps (*Coprinus* spp.) are favoured as edibles in this group. However, some species of both have caused **intestinal upsets** (see mushroom toxins, p. 319) in sensitive individuals and initially only small portions should be eaten, even if the mushrooms are known to be edible. In *Agaricus* (pp. 192–94), the gills are free, a ring is on the stalk, and the gills are pink at first but turn a rich dark purple-brown at maturity.

Almost all the *Coprinus* spp. (pp. 195–99) autolyse (self-digest) at maturity into a black, inky fluid. They have a very short shelf life and should be eaten the day they are picked.

Key to Genera of Dark-Spored Mushrooms

Spores are purple-brown, blackish-brown, black, etc.

1. Gills free from stalk, ring prominent and persistent .... ***Agaricus***
1. Gills attached **2**

2. Gills running down stalk (decurrent) **3**
2. Gills attached but not decurrent **4**

3. Flesh of caps white .... ***Gomphidius***
3. Flesh of caps yellow .... ***Chroogomphus***

4. Caps turning to ink at maturity (deliquescing) .... ***Coprinus***
4. Caps not deliquescing **5**

5. Ring present on stalk .... ***Stropharia, Psathyrella***
5. Ring absent **6**

6. Caps tiny (< 1 cm in diameter), in large clusters, on wood .... ***Coprinus disseminatus***
6. Caps > 1 cm in diameter **7**

7. Caps yellow to yellow-brown, in large clusters on wood .... ***Hypholoma***
7. Caps light to dark brown or grey **8**

8. Gills mottled, caps hemispherical, fruiting in grass .... ***Panaeolina***
8. Gills not mottled, fruiting in grass, on wood, or on dung **9**

9. Fruiting on or near herbivore dung .... ***Panaeolus***
9. Fruiting on wood or in grass **10**

10. Caps hemispherical to convex .... ***Psathyrella***
10. Caps conical to bell-shaped, with a pointed knob .... ***Psilocybe***

Agaricus bitorquis

Sidewalk Mushroom

Caps are 3–12 cm across, convex, becoming flat and sometimes depressed, smooth to silky, dry, and white or with a yellow or ochre tinge. Flesh is white, not staining. Gills are free, close to crowded, narrow, and pink, becoming very dark brown. Stalks are short and stout, up to 5 cm tall by 2.5 cm wide, smooth, and white. Ring is double, white, and persistent. Spore print is purple-brown. Widespread and common, Sidewalk Mushroom favours hard-packed soil. Also known as *A. edulis*. Edible.

Photo: Greg Thorn.

Agaricus silvicola

Caps are 5–12 cm across, convex to flat, dry, smooth, and white, staining yellow when bruised. Flesh is white, staining yellow. Gills are free, crowded, and white to pink, becoming dark brown. Stalks are up to 12 cm tall by 2 cm wide, white, staining yellow, and often with a bulbous base. Ring is persistent, large, floppy, and double. Spore print is dark brown to purple-brown. Widespread and fairly common, this species fruits on the ground in woods and is recognized by its smooth, white, yellow-staining caps. Edible.

Agaricus diminutivus

Caps are 1–4 cm across, convex to flat, dry, and pale brown, with a pinkish to violet-grey tinge, bruising brown. Gills are free, pinkish, becoming dark purple-brown, and close. Stalks are up to 5 cm tall by 8 mm wide and white. Ring is narrow, delicate, and often disappearing. Spore print is purple-brown. Widely distributed and not uncommon, this tiny species fruits on the ground in woods and pastures.

Agaricus campestris

Meadow Mushroom

Caps are 2–8 cm across, convex to flat, white, ageing brown, and sometimes becoming scaly in age or during dry weather. Gills are free, close to crowded, and pink at first, ageing blackish-brown. Stalks are up to 7 cm tall by 2 cm wide and more or less smooth. Ring is narrow, delicate, and often disappearing. Spore print is purple-brown. Widespread, Meadow Mushroom fruits in lawns and pastures. Edible.

Agaricus silvaticus

Wood Mushroom

Caps are 5–10 cm across, convex, dry, scaly, and greyish, becoming reddish. Flesh is white, staining reddish. Gills are free, close, and pink, becoming dark brown. Stalks are up to 9 cm tall by 1.5 cm wide, dry, nearly smooth, and white or staining as the cap. Ring is persistent and white. Spore print is dark brown. The flesh of this scaly mushroom stains red but not as dramatically as Bleeding Mushroom (*A. haemorrhoidarius*, p. 194). Widespread but not common, Wood Mushroom fruits on the ground in woods. Edible.

Photo: Greg Thorn.

Agaricus placomyces

Flat Top

Caps are 5–15 cm across, convex to flat, dry, pallid, and more or less covered with blackish-brown or grey-black scales. Flesh is white to pink, staining yellow to brown. Gills are free, crowded, and pink, becoming purple-brown. Stalks are up to 12 cm tall by 1.5 cm wide, white, staining yellow, and smooth. Ring is large and persistent. Spore print is dark brown. Widespread but not common, Flat Top is recognized by its black scales over a beige base, but this woodland mushroom's appearance can vary widely, depending on the concentration of scales. It fruits on the ground in woods. Also known as *A. praeclaresquamosus*.

Agaricus haemorrhoidarius

Bleeding Mushroom

Caps are 5–10 cm across, convex, becoming flat, dry, grey-brown, and covered with red-brown scales or hairs. Flesh is white, staining red rapidly. Gills are free, close, and pink to purple-brown. Stalks are up to 12 cm tall by 1.5 cm wide, smooth or slightly hairy, and white, becoming pinkish-tan. Ring is large, persistent, and white. Spore print is purple-brown. Widespread but not common, Bleeding Mushroom fruits on the ground in marshy woods. Edible.

Coprinus comatus

Shaggy Mane

Caps are 5–15 cm across, conical to bell-shaped, dry, and covered with brown or blackish, recurved scales. Gills are free, crowded, and white to beige, becoming black and dissolving into a black fluid. Stalks are up to 15 cm tall by 2 cm wide, smooth, and hollow. Ring is white and movable. Spore print is black. Widespread and common, Shaggy Mane fruits in grass and is associated with disturbed sites. Edible.

Coprinus quadrifidus

Caps are 2–8 cm across, conical to bell-shaped, dry, covered with cottony patches that disappear in age, striate, and whitish to grey or tan-grey. Gills are free, close, and white, becoming inky. Stalks are up to 10 cm tall by 1.5 cm wide and white, with a cottony covering. Ring disappears. Spore print is black. Widespread and locally common, this inky cap fruits early in the year on well-rotted wood. Also known as *C. variegatus*. Edible.

Coprinus atramentarius

Tippler's Bane

Caps are 2–8 cm across, conical to bell-shaped, metallic-grey, smooth, and silky-fibrillose. Gills are free, crowded, white, becoming black, and dissolving into a black fluid. Stalks are up to 15 cm tall by 2 cm wide, white, and hollow. Ring disappears early. Spore print is black. Widespread and common, Tippler's Bane fruits in grassy places growing on buried wood. Edible, but do **not** consume with alcohol (see toxins, p. 321). *Coprinus* spp. autolyse very quickly and you must eat them the same day that you pick them!

Coprinus plicatilis

Japanese Parasol

Caps are 1–2.5 cm across, conical to bell-shaped, then parasol-shaped, dry, thin, with deep, radial grooves and a scalloped margin, and brown to grey-black. Gills are far apart and narrow. Stalks are up to 7 cm tall by 2 mm wide and smooth. Spore print is black. Widespread and locally common, this attractive inky cap fruits in grass beside roads and pathways or on well-rotted wood.

Coprinus micaceus

Mica Cap

Caps are 1–5 cm across, conical to bell-shaped, striate, dry, tan to ochre, and covered with tiny, glistening particles that disappear. Gills are attached, crowded, and white, becoming inky. Stalks are up to 8 cm tall by 6 mm wide, white, silky, and hollow. Spore print is black. Widespread and common, Mica Cap fruits in large numbers on or near stumps, often in grass, from buried wood or dead roots. Edible.

Coprinus disseminatus

Crumble Cap

Caps are 0.5–1 cm across, conical to bell-shaped, dry, blue-grey or grey-brown, and striate. Gills are attached and white, becoming black. Stalks are up to 4 cm tall, slender, white, and minutely hairy, becoming smooth. Spore print is black to purple-black. Widespread and common, tiny Crumble Cap fruits in large numbers on old stumps or from buried wood. It is one of the few species in this group that doesn't turn into ink.

Coprinus lagopus

Caps are 1–3 cm across, ellipsoid, becoming bullet-shaped and then flat, with an upturned margin, grey, flecked with brown near the centre, and covered with loose, fibrous scales. (Variations of the cap shapes are shown in the photos.) Gills are free, narrow, and grey, becoming black. Stalks are up to 10 cm tall, slender, white, and covered with dense, woolly fuzz. Spore print is black. Widespread, this species is the most common of a bunch of small inky caps that can only be distinguished microscopically.

Coprinus radians

Caps are 2–4 cm across, egg-shaped, becoming conical to narrowly bell-shaped, tan to ochre, scaly to flaky, becoming smooth, and striate. Gills are close and white, becoming black and turning to ink. Stalks are up to 8 cm tall by 1 cm wide, arising from an orange-red to red-brown fungal mat. Spore print is black. Widespread but not common, this species fruits on the debris of hardwood trees.

Chroogomphus rutilus

Caps are 5–12 cm across, convex to broadly convex, with a small knob, smooth, olive-brown to red-brown, and slimy when wet. Flesh is yellowish to salmon. Gills are decurrent, well-spaced, and buff to olive-purple. Stalks are 15 cm tall by 2.5 cm wide, fibre-streaked, and orange, with reddish tints. Spore print is smoky-brown to black. Widespread and not uncommon, this distinctive species fruits under conifers. Edible.

Chroogomphus vinicolor

Caps are 2–7 cm across, convex, knobbed, dark red-brown to vinaceous, with purplish tints, and slimy when wet. Gills are decurrent, well-spaced, and coloured as the cap. Stalks are up to 8 cm tall by 12 mm wide and buff to ochre, ageing or staining purplish. Spore print is smoky-grey to blackish. Widespread and not uncommon, this species fruits under conifers. Edible.

Gomphidius glutinosus

Caps are 5–10 cm across, convex to flat, becoming depressed, slimy when wet, smooth, and grey-brown to purple-brown. Gills are decurrent, white, becoming grey-brown, and far apart. Stalks are up to 9 cm tall by 2 cm wide, smooth or slightly hairy, and white to tan, with a yellow base. Ring is slimy. Spore print is blackish-grey. Widespread but not common, this species fruits under conifers. Edible.

Gomphidius subroseus

Rosy Gomphidius

Caps are 4–7 cm across, convex, becoming flat and depressed, smooth, slimy, and rose-pink to salmon-red. Gills are decurrent, well-spaced, and white, becoming grey-brown. Stalks are up to 7 cm tall by 2 cm wide and white, with a yellow base. Ring is white, becoming blackish with the spore deposit, sometimes disappearing. Spore print is black. Widespread but rare in Ontario, eastern Canada and the adjacent U.S., this attractive and distinctive mushroom fruits under conifers. Edible.

Hypholoma sublateritium

Brick Top

Caps are 2–10 cm across, broadly convex, sometimes knobbed, smooth, moist to dry, and brick-red, paler near the margin. Gills are attached, close, narrow, and white to grey, becoming purple-brown. Stalks are up to 10 cm tall by 12 mm wide and whitish. Spore print is purple-brown. Widespread and common, Brick Top fruits on dead hardwood. Edible, but very similar to **poisonous** Sulphur Tuft (below).

Hypholoma fasciculare

Sulphur Tuft

Caps are 2–10 cm across, convex to flat, sometimes with a central knob, smooth, dry, orange-yellow to sulphur-yellow, darker at the centre, and with a bitter taste. Gills are attached, close, and yellow to greenish-yellow, becoming purple-brown. Stalks are up to 8 cm tall by 1 cm wide, yellow, staining brown, and smooth to slightly scaly. Spore print is purple-brown. Widespread but probably not as common in the east as generally believed, Sulphur Tuft fruits in clusters on logs and stumps of conifers. **Poisonous.**

Hypholoma dispersum

Dispersed Hypholoma

Caps are 2–4 cm across, convex to bell-shaped, smooth to silky-striate, and red-brown to ochre. Inner veil can attach to the cap. Gills are attached, crowded, and white to greyish, becoming purple-brown. Stalks are up to 10 cm tall by 6 mm wide and mottled, with whitish, hairy patches. Ring disappears. Spore print is purple-brown. Widespread and locally common, Dispersed Hypholoma fruits under conifers. Also known as *H. marginatum*.

Hypholoma capnoides

Caps are 2–7 cm across, convex to flat, smooth, moist to dry, reddish-brown to ochre, paler at the margin, and with the inner veil remnants attached to the margin. Gills are attached, close, and white at first, becoming purple-brown. Stalks are up to 8 cm tall by 1 cm wide, white to tan, and brown at the base. Spore print is purple-brown. Widespread and common, this species fruits in dense clusters on conifer wood. Edible, but don't confuse it with **poisonous** Sulphur Tuft (p. 201).

Psathyrella velutina

Weeping Widow

Caps are 3–8 cm across, hemispherical to flat, dry, tan to ochre, and covered with dark brown hairs. Gills are attached to free, crowded, white to purple-black, and mottled, often with droplets of fluid at the edge of the gills. Stalks are up to 12 cm tall by 1 cm wide and fibrous to scaly. Spore print is purple-brown to blackish. Widespread and common, Weeping Widow fruits in grass or in open woods. Edible, but **not** recommended.

Psathyrella piluliformis

Caps are 2–5 cm across, convex to flat, slightly knobbed, smooth or silky, and tan to dark brown when wet, drying buff. Gills are attached, close, narrow, and greyish to purple-brown. Stalks are up to 8 cm tall by 6 mm wide, white, and smooth but powdery above. Spore print is purple-brown. Common and widespread, this species fruits on well-rotted wood. Also known as *P. hydrophila*.

Psathyrella candolleana

Caps are 2–5 cm across, convex to bell-shaped, becoming knobbed, dry, smooth, honey-brown when wet, and drying cream to white. Margin is often adorned with remnants of the inner veil. Gills are attached, close, narrow, and white, becoming purple-brown. Stalks are up to 8 cm tall by 6 mm wide, smooth, and white. Ring disappears. Spore print is purple-brown. Widespread and fairly common, this species fruits on or near well-rotted wood. Edible.

Psathyrella multipedata

Caps are 1–3 cm across, bluntly conical to convex, clay-brown to yellow-brown, ageing dark brown to blackish-brown, faintly striate, and with a bitter taste. Gills are attached, close, and tan, ageing purple-brown. Spore print is blackish-brown. Stalks are up to 12 cm tall, slender, and white. Widespread but not common, this species fruits on the ground in dense clusters in open woods, often beside trails.

Psathyrella conopilea

Bell-Shaped Psathyrella

Caps are up to 4 cm across, conical to broadly bell-shaped, chocolate-brown to grey-brown, fading to tan, and radially striate. Gills are attached, close, and pale brown, becoming blackish-brown. Stalks are up to 12 cm tall, slender, and white. Spore print is blackish-brown. Widespread but not common, Bell-Shaped Psathyrella fruits on rich soil, often in grassy places.

Photo: Greg Thorn.

Panaeolina foenisecii

Haymaker's Mushroom

Caps are 1–3 cm across, convex to bell-shaped, dry, smooth or cracking in age, and grey-brown to smoky- or red-brown, drying buff. Gills are attached, close, broad, dark brown to purple-black, and mottled. Stalks are up to 8 cm tall, slender, smooth, and pale brown. Spore print is purple-brown. Widespread, Haymaker's Mushroom is the commonest of some related little brown mushrooms (LBMs). Also known as *Panaeolus foenisecii*, this species fruits on lawns. **Poisonous, but not deadly** (see p. 322).

Psilocybe semilanceata

Liberty Cap

Caps are up to 2 cm across, narrowly conical to bell-shaped, with a sharply pointed knob, sticky, chestnut-brown to yellow-brown, and darker at the apex, often fading at the margin. Gills are attached and pale brown, becoming blackish-brown. Stalks are up to 8 cm tall, slender, and white to pale tan. Spore print is purple-brown. Widespread, Liberty Cap is rare in Ontario but common in the Maritime provinces, where it fruits in pastures. Also found in the eastern U.S. **Poisonous** (**hallucinogenic**, see p. 322).

Photo: Greg Thorn.

Panaeolus sphinctrinus

Bell-Shaped Panaeolus

Caps are 2–5 cm across, conical, becoming bell-shaped, smooth, moist to dry, with a toothed margin, grey-brown to grey-olive, with a satin sheen, and sometimes with the inner veil remnants clinging to the margin. Gills are attached, well-spaced, broad, and grey, becoming mottled, with white edges. Stalks are up to 12 cm tall, slender, and coloured as the cap. Spore print is black. Widespread and common, Bell-Shaped Panaeolus fruits on or near herbivore dung in woods and pastures.

Psathyrella delineata

Caps are 4–10 cm across, broadly conical, becoming bell-shaped and broadly knobbed, dry, wrinkled at the centre, and rich red-brown when young, maturing to orange-brown, then fading in age. Stalks are up to 10 cm tall by 2 cm wide, whitish, darker near the base, and fibrous. Ring is fragile, grooved, and disappears. Spore print is purple-brown. Widespread but not common, this handsome species fruits on woody debris.

Stropharia aeruginosa

Blue-Green Stropharia

Caps are 2–5 cm across, convex, becoming flat, knobbed, slippery when wet, bright blue-green to dull blue-green, fading to yellowish, and sometimes with white scales at the margin. Gills are attached, close, and whitish to smoky, ageing purple-brown. Stalks are up to 7 cm tall by 9 mm wide, sticky, scaly, whitish to blue-green, and hollow. Ring is persistent and narrow. Spore print is purple-brown. This distinctive species is widespread but not common. It fruits on the ground in woods.

Stropharia thrausta

Caps are 3–8 cm across, hemispherical to broadly convex, knobbed, with cottony scales, especially near the margin, slippery when wet, and orange-red to brick-red. Gills are attached, close, and whitish, ageing purple-brown. Stalks are up to 10 cm tall by 1 cm wide. Spore print is purple-brown. This distinctive species is widespread but not common, and it fruits on the ground in mixed woods. **Poisonous.**

Photo: Greg Thorn.

Stropharia hornemannii

Caps are 2–13 cm across, convex to flat, with a cottony margin, sticky, slippery when wet, smooth, and brown to red-brown, sometimes with a purplish tinge. Taste is unpleasant. Gills are attached, close, and pale grey, becoming purple-brown. Stalks are up to 15 cm tall by 2 cm wide, white to yellow, and shaggy below the ring. Ring is persistent and white to brown. Spore print is purple-brown. Widespread but not common, this species fruits under conifers.

Stropharia semiglobata

Caps are 1–4 cm across, hemispherical, yellowish to tan, slippery when wet, and smooth. Gills are attached, close, and grey to purple-brown. Stalks are up to 12 cm tall, slender, and white to pale yellow-tan. Ring is white, turning black with spores. Spore print is purple-brown. Widespread and fairly common, this species fruits on dung and is recognized by the combination of this fruiting habit, the cap shape, and the black spore print.

MUSHROOMS WITH BROWN SPORE PRINTS

Spore prints of mushrooms in this group are tan, clay, ochre, brown or rusty-brown but not dark brown. Assessment of the colour of spore prints is subjective. It will vary according to the thickness of the deposit and whether the print is fresh or dry. In a number of genera, the spores are distinctly red-brown and the rust-coloured spore print is an important feature for identification. Initially, you often make a guess on the spore colour by checking the ring on the stalk or the caps of adjacent mushrooms or plants in the vicinity. A preliminary assessment can be confirmed later by a *bona fide* spore print on white paper. It always makes sense, when in doubt, to key your specimen out in both sections of the key, i.e., "rusty-spored" and "non-rusty-spored."

There are many interesting fungi in this group, but in general brown-spored mushrooms are **not highly regarded** as a source of edible fungi. *Pholiota* spp. (pp. 215–18) are fairly common. They are brown-spored, bright yellow to yellow-brown and often scaly, and they have a ring. For the most part, *Pholiota* spp. are wood-rotters and are found commonly on logs or stumps, but some species fruit on the ground from dead roots, buried wood and the like. Although often found in impressive fruitings, *Pholiota* spp. are not generally favoured as edibles. *Pholiota squarrosa* is enjoyed by some, but this same species causes **gastrointestinal distress** in others. Also, Deadly Galerina (*Galerina autumnalis*, p. 222) is very similar to the edible Changeable Pholiota (*Pholiota mutabilis*, p. 218) and mistakes are too easy to make! The Gypsy (*Rozites caperata*, p. 219) is a Pholiota-like species that is considered a good edible. It has a persistent ring and attached gills, but the spores are rust-brown and it is solitary or scattered on the ground and never in clumps. Sulphur Tuft (*Hypholoma fasciculare*, p. 201) and related species grow in clusters on wood and can be confused with *Pholiota* by beginners, but they have a dark purple-brown spore print.

Cortinarius (pp. 226–30) is one of the most frequently encountered of the rusty-spored genera, and is probably one of the commonest of all the woodland fungi. Instead of a membranous inner veil, which would form the ring on the stalk, *Cortinarius* spp. have a cobweb-like veil protecting the developing gills. This veil is seen best in young specimens. As the cap expands, the cobweb veil disappears and, at best, survives as a stain on the stalk, so it is easily overlooked. With a little practice, it is easy enough to recognize a *Cortinarius* but identification to species is another matter. There are hundreds of described species in this genus, and these are often very similar in appearance. In this book, I have illustrated some *Cortinarius* spp. that are common and distinctive so that they can be identified with reasonable confidence. Most of the species of *Cortinarius* found will be unidentifiable. While some are regarded as edible and good (e.g., *C. violaceus*), other species are known to contain the **deadly** toxin orellanine (p. 320), so it is hard to be enthusiastic about recommending **any** *Cortinarius* spp. as edible. Some of the smaller, brightly coloured *Cortinarius* spp. have been separated into the genus *Dermocybe*.

This group contains a large number of little brown mushrooms (LBMs) belonging to *Conocybe*, *Inocybe*, *Agrocybe*, etc., that are known to contain **toxins** of one type or another. All LBMs should be **avoided** as edibles.

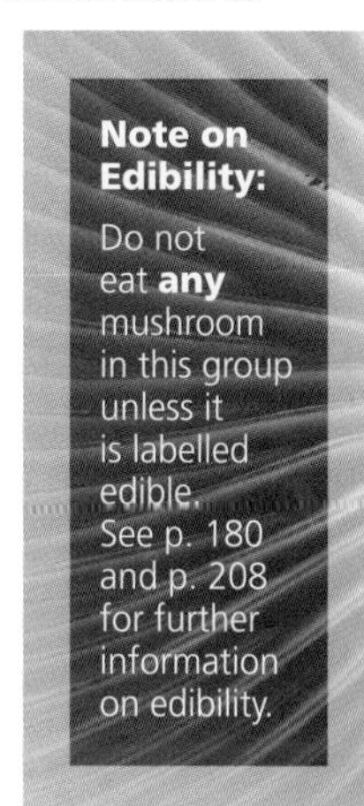

Key to Genera of Brown-Spored Mushrooms

FB = fruitbodies

1. Gills reduced to ridges running down the stalk .... ***Gomphus***
1. Gills well developed **2**

2. Stalks lateral or absent **3**
2. Stalks central **4**

3. Caps fleshy, gills running down stalk (decurrent) .... ***Paxillus***
3. Caps thin, shelving, gills not decurrent .... ***Tapinella, Crepidotus***

4. Spores rust-brown **5**
4. Spores ochre, clay-coloured, or brown **13**

5. Caps average (< 3 cm in diameter), conical to convex or bell-shaped with a slender stalk **6**
5. Caps average (> 3 cm in diameter), convex **8**

6. Fruiting on ground or in grass **7**
6. Fruiting in moss, on wood, or herbivore dung .... ***Pholiota, Galerina***

7. Stalks very long, white, curving over .... ***Gastrocybe***
7. Stalks erect .... ***Conocybe, Bolbitius***

8. Ring not present **9**
8. Membranous or cobweb ring present (check young FB) **10**

9. Gills decurrent .... ***Omphalotus***
9. Gills not decurrent .... ***Pholiota, Gymnopilus***

10. Ring cobweb-like, disappearing .... ***Hebeloma, Cortinarius***
10. Ring membranous, persistent **11**

11. Fruiting on the ground, cap radially wrinkled .... ***Rozites***
11. Fruiting on wood **12**

12. Caps large (5–15 cm in diameter) .... ***Gymnopilus***
12. Caps smaller (2–6 cm in diameter) .... ***Galerina autumnalis, Pholiota mutabilis***

13. Growing on old mushrooms .... ***Nyctalis***
13. Growing on ground or wood, etc. **14**

14. Gills running down stem (decurrent) **15**
14. Gills not decurrent **16**

15. Gills bright yellow, well-separated, bruising blue .... see ***Phylloporus***
15. Gills close .... ***Paxillus***

16. Ring present (cobweb or membrane) **17**
16. Ring absent **19**

17. Stalks with cobweb-like ring .... ***Hebeloma***
17. Stalks with membranous ring **18**

18. On ground in grass .... ***Agrocybe***
18. On wood or wood chips .... ***Tubaria, Pholiota***

19. Fruits on ground or dung **20**
19. Fruits on wood or woody debris **23**

20. Fruits in lawns, pastures **21**
20. Fruits in woods **22**

21. Caps slimy, yellowish, on manured soil, in grass .... ***Bolbitius***
21. Caps dry, cream to tan or yellow-brown .... ***Agrocybe***

22. Stalks rooting, cap slimy .... ***Phaeocollybia***
22. Stalks not rooting, cap dry .... ***Agrocybe, Inocybe***

23. Caps brown, solitary or few .... ***Tubaria***
23. Caps yellow to yellow-brown, in clusters .... ***Pholiota, Hypholoma***

Bolbitius vitellinus

Caps are 2–5 cm across, conical to bell-shaped, becoming flat and knobbed, slippery when wet, smooth, and yellow to mustard-yellow. Gills are attached, narrow, and dull ochre, becoming rusty. Stalks are up to 10 cm tall by 8 mm wide, powdery or with scattered hairs, and white or with a yellow tinge. Spore print is rust-brown. Widespread and not uncommon, this species fruits on rich soil or well-rotted wood. Edible.

Conocybe filaris

Deadly Conocybe

Caps are 5–15 mm across, hemispherical to bell-shaped, and tan to yellow-brown, with a striate margin. Gills are attached and pale brown to ochre. Stalks are up to 5 cm tall by 3 mm wide, white to yellowish-brown, with a prominent, membranous, striate ring. Spore print is rust-brown. Widespread and not uncommon in grassy spots in woods, Deadly Conocybe often fruits beside trails. **Deadly poisonous**.

Conocybe tenera

Caps are 8–20 mm across, conical to bell-shaped, dry, smooth, striate, and pale red-brown, fading to whitish. Gills are attached to nearly free and pallid, becoming rusty. Stalks are up to 7 cm tall, slender, striate, and coloured as the cap. Spore print is rust-brown. Widespread but not common, this species fruits in grassy spots.

Photo: Greg Thorn.

Agrocybe vervacti

Caps are 2–4 cm across, hemispherical to broadly convex, ochre to orange-brown, dry, and smooth. Gills are attached, close, and brown. Stalks are up to 7 cm tall by 8 mm wide, beige to pale orange-brown, slightly roughened, and striate (with darker fibrils). Spore print is brown. Common in the Great Lakes region, this species fruits in grass on lawns in summer.

Gastrocybe lateritia

Caps are 1–1.5 cm across, narrowly conical, slimy, smooth, off-white to tan or yellowish, and striate. Gills are free, narrow, ochre, and quickly gelatinizing and clinging to the stalk. Stalks are up to 12 cm tall by 2 mm wide, smooth, hollow, white, and curving markedly, which, in part, causes the cap to stick to grass blades. No spore print because gills gelatinize early. Widespread and locally common around the Great Lakes and probably farther afield, this distinctive mushroom fruits in grass in early summer.

Conocybe lactea

Dunce Cap

Caps are 1–3 cm across, conical to bell-shaped, dry to moist, smooth or radially wrinkled, and tan to cream or almost white. Flesh is thin and white. Gills are attached and whitish, becoming rusty. Stalks are up to 10 cm tall, slender, and minutely hairy. Spore print is rust-brown. Common and widespread, Dunce Cap fruits on lawns and is recognized by its pale, conical cap.

Agrocybe erebia

Caps are 2–5 cm across, convex, becoming flat, with a low knob, smooth, and brown to blackish-brown, fading to tan. Gills are attached and off-white, ageing rusty. Stalks are up to 5 cm tall by 7 mm wide, white, and striate. Ring is large, persistent, white, and striate above. Spore print is brown. Widespread but uncommon, this species is recognized by its colour and large, persistent ring. It fruits on the ground in woods.

Agrocybe acericola

Maple Agrocybe

Caps are 2–8 cm across, convex, becoming flat, with a broad knob, dry, smooth, radially wrinkled, and buff to yellow-brown. Gills are attached, broad, close, and off-white, ageing brown. Stalks are up to 10 cm tall by 1 cm wide and striate. Ring is large, white, persistent, and striate on the upper surface. Spore print is dull brown. Widespread but not common, Maple Agrocybe fruits on rotten hardwood or on the ground. Edible, but **not** recommended (see p. 208).

Photo: Greg Thorn.

Agrocybe molesta

Cracked-Top

Caps are 2–5 cm across, hemispherical to broadly convex, dry, smooth, becoming cracked and fissured in age, and creamy white to yellowish. Gills are attached, close, and off-white to dark brown. Stalks are up to 8 cm tall by 1 cm wide and coloured as the cap. Ring is small, sometimes disappearing. Spore print is brown. Widespread and common, Cracked-Top fruits in grass. It is also known as *A. dura*. Edible, but **not** recommended (see p. 208).

Phaeocollybia christinae

Caps are 1–4 cm across, conical to broadly conical, with a sharply pointed knob, sticky, smooth, and ochre to red-brown. Gills are attached and whitish, becoming pale brown, ageing red-brown, and with rusty stains. Stalks are up to 14 cm tall by 6 mm wide, with half the length or more rooted in the substrate, smooth, brittle, and coloured as the cap to rich red-brown. Spore print is brown. Widespread but not common, this species fruits in debris below conifers or in mixed woods, often in moss. Identified by its rooting habit.

Pholiota aurivella

Golden Pholiota

Caps are 3–15 cm across, hemispherical to convex, becoming flat, yellow to yellow-brown, with red-brown scales, and slimy. The fruitbodies persist for some time and the scales might weather off. Gills are attached, broad, close, and yellow, becoming rusty. Stalks are up to 10 cm tall by 12 mm wide, coloured as the cap, and scaly. Ring is inconspicuous. Spore print is brown. Widespread and common, Golden Pholiota fruits on dead wood. Edible, but **not** recommended (see p. 208).

Pholiota flammans

Flaming Pholiota

Caps are 4–8 cm across, convex, becoming flat, dry, covered with yellow, cottony scales, and bright golden-yellow to orange-yellow. Gills are broadly attached, close, and yellow, becoming rust-brown. Stalks are up to 10 cm tall by 1 cm wide, bright yellow, and scaly. Ring is yellow and disappearing. Spore print is brown. Widespread but not common and recognized by its brilliant colour, Flaming Pholiota fruits on conifer logs and stumps.

Pholiota squarrosa

Scaly Pholiota

Caps are 2–12 cm across, hemispherical to broadly convex, dry, and yellow-brown, with darker, recurved scales. Gills are attached and yellowish, with greenish tints, becoming olive-brown. Stalks are up to 12 cm tall by 18 mm wide and white to yellowish, with brownish scales. Ring is persistent or disappearing. Spore print is brown. Common and widespread, Scaly Pholiota is recognized by its dry, scaly cap and that it fruits in clusters at the base of tree trunks.

Pholiota squarrosoides

Caps are 2–10 cm across, convex to flat, with a low knob, slimy when wet, and densely scaly, with tan scales on a white to buff background. Gills are attached, close, and whitish, becoming brown. Stalks are up to 10 cm tall by 1 cm wide and scaly. Ring is whitish and disappearing. Spore print is brown. Widespread and not uncommon, this species is recognized by its slimy and scaly cap and that it fruits in clusters on or near hardwood trees or stumps. Edible.

Pholiota malicola

Caps are 3–10 cm across, convex, becoming flat, broadly knobbed, smooth, and yellow, becoming shiny. Gills are attached, yellowish, becoming brown, and close. Stalks are up to 12 cm tall by 15 mm wide, coloured as the cap, roughened by the fibrils of the inner veil remnants below the ring, and red-brown at the base. Ring is disappearing. Spore print is reddish-brown. Widespread and not uncommon, this species fruits in clusters on or near stumps.

Pholiota granulosa

Caps are 1–3.5 cm across, convex to broadly convex, dry, with a yellowish-brown base, covered with tiny, dark brown scales, and with remnants of the yellowish inner veil at or near the margin. Gills are attached, close, yellow, becoming brown, and spotted. Stalks are up to 5 cm tall by 3 mm wide, scaly and flaky, and pale yellow, darker below. Ring disappears. Widespread and not uncommon, this small, distinctive species fruits on the logs and stumps of hardwoods and conifers.

Pholiota mutabilis

Changeable Pholiota

Caps are 2–6 cm across, convex to flat, knobbed, with a striate margin, and orange-brown, fading to clay-brown, drying paler. Gills are attached, close, and whitish, ageing brown. Stalks are up to 10 cm tall by 12 mm wide, coloured as the cap, dark brown, and scaly below the ring. Ring disappears. Spore print is ochre to brown. Widespread, Changeable Pholiota fruits in clusters on stumps and logs of conifers and hardwoods. Can be confused with Deadly Galerina (*Galerina autumnalis*, p. 222). Also known as *Kuehneromyces mutabilis*. Edible, but **not** recommended (see p. 208).

Pholiota astragalina

Bitter Pholiota

Caps are 2–5 cm across, convex to bell-shaped, becoming broadly knobbed, bright apricot-orange, sometimes bruising black, sticky, and with a bitter taste. Gills are attached, yellow-orange, becoming brown, and close. Stalks are up to 10 cm tall by 5 mm wide, yellowish, and smooth. Ring is yellowish and disappearing. Spore print is brown. Widespread but not common in Ontario, eastern Canada and the adjacent U.S., Bitter Pholiota fruits on conifer stumps and logs.

Photo: Libby Fox.

Rozites caperata

The Gypsy

Caps are 5–10 cm across, convex to flat or knobbed, straw-coloured to ochre, dry, and with radial wrinkles. Gills are attached and whitish, becoming rust-brown. Stalks are up to 12 cm tall by 2 cm wide, smooth to slightly fibrillose, and whitish. Ring is whitish and persistent. Spore print is rust-brown. Widespread and common, The Gypsy fruits on the ground in woods. Edible.

Tubaria furfuracea

Caps are 1–3 cm across, convex to flat, tan-brown, dry, smooth or with radial fibres, and sometimes with inner veil remnants sticking around the margin. Gills are attached and coloured as the cap. Stalks are up to 4 cm tall by 4 mm wide and coloured as the cap. Spore print is ochre. Widespread and not uncommon, this species fruits during wet periods on sticks or wood chips in spring or early summer.

Photo: Greg Thorn.

Tubaria confragosa

Caps are 2–4 cm across, bell-shaped to convex, red-brown to pale tan, covered with white, floccose material, thin, fragile, and with a striate margin. Stalks are up to 5 cm tall by 3 mm wide, with a persistent ring. This brown-spored species is recognized by the fuzzy, white growth over its cap and stalk. It is widespread but not common and fruits sparingly on logs and stumps. It is also known as *Pholiota confragosa*.

Crepidotus applanatus

Flat Crep

Caps are 1–3 cm across, convex to flat, shell-shaped to fan-shaped, shelving, dry, smooth, and white. Gills are attached, narrow, close, and white, ageing brown. Stalks are absent or very short, with a hairy base. Spore print is brown. Widespread and common, Flat Crep fruits on rotting logs and stumps.

Crepidotus mollis

Caps are 1–7 cm across, semi-circular in outline, shelving, laterally attached, dry, and covered with tan to red-brown scales that disappear in older specimens. Gills are close and whitish, becoming brown. Stalks are absent. Spore print is dull brown. Widespread and common, this species fruits on decaying hardwood logs and stumps.

Crepidotus variabilis

Caps are 1–2 cm across, fan-shaped to kidney-shaped, with a wavy, inrolled margin, shelving, laterally attached, downy to velvety, soft, and fragile. Gills are well-spaced and white, ageing brown. Stalks are absent. Spore print is brown. Widespread and not uncommon, this species fruits in clusters on hardwood twigs. There are a number of small *Crepidotus* spp. that can only be distinguished microscopically.

Crepidotus crocophyllus

Caps are 1–4 cm across, kidney-shaped to fan-shaped, orange to orange-brown, and minutely scaly. Gills are narrow, close, and coloured as the cap. Stalks are absent. Spore print is brown. Widespread but not common, this species fruits on hardwood logs and stumps, and is easily recognized by its brilliant colour. The brown spores separate it from the similar pink-spored *Phyllotopsis nidulans*.

Tapinella panuoides

Caps are 3–12 cm long, up to 5 cm wide, bracket-like or fan-shaped, with a wavy margin, dry, downy, becoming smooth, and ochre to olive-yellow. Gills are whitish to ochre and forked. Spore print is buff. Widespread but uncommon, this species fruits on conifer stumps. Also known as *Paxillus panuoides*.

Galerina paludosa

Caps are 1–3 cm across, conical to convex or bell-shaped, knobbed, smooth, and tan. Gills are close, attached, and tan. Stalks are up to 12 cm tall, slender, and coated in intervals with white patches. Spore print is rusty-brown. Widespread and common in mossy bogs, this species fruits early in the year and is recognized by its patchy stalk.

Gymnopilus spectabilis

Big Laughing Mushroom

Caps are 5–12 cm across, convex to flat, yellow-brown to orange-brown, dry, and smooth. Flesh is yellowish and bitter-tasting. Gills are attached, close, and yellowish, ageing rusty. Stalks are up to 15 cm tall and 2 cm wide, coloured as the cap, and fibril-streaked below the ring. Ring is persistent and yellow-brown, becoming dark brown with spores. Spore print is reddish-brown. Widespread and fairly common, Big Laughing Mushroom fruits in clusters on dead wood. **Poisonous.**

Galerina autumnalis

Deadly Galerina

Caps are 2–6 cm across, convex, moist, yellow-brown to buff, and with a striate margin. Gills are attached, tan to rust-brown, and close. Stalks are up to 9 cm tall by 8 mm wide, smooth, pale brown, and dark brown near the apex, with a membranous ring. Spore print is rust-brown. Widespread and common, Deadly Galerina fruits in clusters on wood, usually late in the year. Note the membranous ring and smooth stalk! **Deadly poisonous.**

Gymnopilus luteofolius

Caps are 2–8 cm across, convex, becoming broadly knobbed, dry, red-brown, paler and yellowish at the centre, and covered with darker spines, especially near the centre. Margin is inrolled and overhanging the gills. Gills are attached, well-spaced, and yellow, becoming reddish-brown. Stalks are up to 10 cm tall by 1.5 cm wide, striate, and coloured as the cap, below a layer of whitish fibrils. Spore print is rusty-orange. Widespread but not common, this species fruits on wood chips, sawdust, rotting logs, etc. **Poisonous.**

Gymnopilus sapineus

Caps are 3–9 cm across, convex to broadly convex, yellow-ochre to orange-brown, minutely hairy, and with a bitter taste. Gills are attached, close, and yellow, becoming brown. Stalks are up to 8 cm tall by 12 mm wide and ochre, darker below. Ring is yellowish, disappearing early. Spore print is orange-brown to rust-brown. Common and widespread, this species fruits on dead conifers.

Photo: Greg Thorn.

Paxillus involutus

Poisonous Paxillus

Caps are 4–15 cm across, convex to flat or depressed, with an inrolled margin, dry, hairy, becoming smooth, and ochre to olive-brown, spotted darker. Gills are decurrent and yellowish, staining brown to red-brown. Stalks are up to 10 cm tall by 3 cm wide, smooth, and coloured as the cap. Spore print is yellow-brown. Common and widespread, Poisonous Paxillus fruits on the ground in woods. **Poisonous**.

Paxillus atrotomentosus

Velvet-Stalked Paxillus

Caps are 5–15 cm across, convex to flat or depressed, dry, and brown to blackish-brown. Margin is inrolled. Gills are decurrent, separable from the cap, close, narrow, forked, and yellow to rusty. Stalks are off-centre, up to 10 cm tall by 4 cm wide, straight or curved, velvety, and blackish, covered with dark brown hairs. Spore print is yellow-brown. Widespread and common, Velvet-Stalked Paxillus fruits on conifer stumps or buried wood.

Hebeloma crustuliniforme

Poison Pie

Caps are 4–6 cm across, convex to flat, with a low knob, dry to slippery when wet, smooth, brown, red-brown at the centre, and fading to pallid at the margin. Flesh is white and bitter-tasting. Gills are attached, close, white to cinnamon-brown, with a toothed edge, and beaded with droplets in humid weather. Stalks are up to 8 cm tall by 15 mm wide, cottony-hairy, and white. Ring is absent. Spore print is ochre to pale brown. Widespread and fairly common, Poison Pie fruits on the ground in open woods. **Poisonous**.

Photo: Greg Thorn.

Inocybe geophylla

Caps are 1–4 cm across, conical to bell-shaped, becoming flat, with a small knob, dry, fibrillose-silky, white to lilac or purple, and splitting at the margin. Gills are attached, close, and white, becoming clay-brown. Stalks are up to 5 cm tall, slender, silky-fibrillose, and white. Spore print is pale brown. Some varieties are white, others are violet, and the collection (right) is bright purple. Widespread and common, this species fruits on the ground in woods. **Poisonous**.

Inocybe fastigiata

Fibrehead

Caps are 2–7 cm across, conical to broadly bell-shaped, knobbed, tan to ochre or dull yellow, dry, and striate with radial fibres and fissures. Gills are attached, close, and grey to olive-brown. Stalks are up to 8 cm tall by 1 cm wide, hairy, and white to smoky-brown. Spore print is dull yellow-brown. Widespread and common, Fibrehead fruits on the ground in woods. **Poisonous**.

Cortinarius alboviolaceus

Caps are 3–10 cm across, bell-shaped to convex, dry, smooth, and pale violet, fading to white. Flesh is pale violet. Gills are attached, close, and pale violet, becoming brown. Stalks are up to 8 cm tall by 2 cm wide, with a clavate base, and web-like inner veil, which forms a hairy ring. Spore print is rust-brown. Widespread and one of the most common species of *Cortinarius*, it fruits in mixed woods.

Cortinarius traganus

Caps are 4–12 cm across, convex, becoming flat, with a down-turned margin, broadly knobbed, silky, dry, and pale violet. Gills are attached, thick, well-spaced, and lilac, becoming red-brown. Stalks are up to 12 cm tall and stout, with a bulbous base. Spore print is red-brown. Widespread and fairly common, this species fruits under conifers. **Poisonous.**

Cortinarius armillatus

Banded Cort

Caps are 5–12 cm across, convex to flat, moist, fibre-streaked, and red-brown to brick-red. Flesh is off-white. Gills are attached, close, and pale brown to rusty. Stalks are up to 15 cm tall by 2 cm or more wide and tan, with several orange-red bands and a swollen, clavate base. Ring is web-like. Spore print is brown. Widespread and common, Banded Cort fruits in conifer woods.

Cortinarius collinitus

Caps are 3–8 cm across, convex to broadly convex, slippery when wet, smooth, and ochre to orange-brown but violet at the margin. Gills are attached, close, and pale violet, ageing red-brown. Stalks are up to 10 cm tall by 2 cm wide, white, stained, with concentric, rusty patches at the base, and slimy. Spore print is rust-brown. Widespread and common, this species fruits on the ground in woods.

Cortinarius corrugatus

Caps are 5–10 cm across, bell-shaped to broadly convex, radially wrinkled, and tan to ochre. Gills are attached and pallid to rusty-cinnamon. Stalks are tan-yellow and up to 12 cm tall by 1.5 cm wide, with a basal bulb. Ring is web-like, leaving a slimy zone on the stalk. Spore print is rust-brown. Widespread and not uncommon, this species fruits under hardwoods.

Cortinarius violaceus

Purple Cort

Caps are 5–12 cm across, convex, becoming flat, with a low knob, dry, covered with minute tufts, becoming shiny in age, and deep violet to purplish-black. Flesh is blue-grey to deep violet. Gills are attached, broad, close to well-spaced, and deep violet. Stalks are up to 12 cm tall by 1.5 cm wide and coloured as the cap. Spore print is rust-brown. Widespread and common, distinctive Purple Cort fruits under conifers. Edible.

Cortinarius iodes

Caps are 2–6 cm across, bell-shaped to convex, slimy, and purple, becoming paler in age. Gills are attached, broad, close, and violet to grey, becoming cinnamon. Stalks are up to 7 cm tall by 1.5 cm wide and slimy. Ring is violet and web-like. Spore print is rust-brown. Widespread and common, this species fruits under hardwoods.

Cortinarius gentilis

Deadly Cort

Caps are 1–5 cm across, conical, becoming flat, with a narrow knob, yellow-brown to brown, smooth, and dry. Gills are attached and yellowish, becoming red-brown. Stalks are up to 10 cm tall, slender, and brown, with yellow remnants of the cobweb inner veil. Spore print is rust-brown. Widespread but not common, small Deadly Cort fruits under conifers in needles or in mossy spots. **Poisonous.**

Cortinarius semisanguineus

Red-Gilled Cort

Caps are 2–7 cm across, convex or bell-shaped to flat, with a low knob, dry, silky to scaly, and yellow-brown. Flesh is white to yellowish. Gills are attached, close, and red to cinnabar. Stalks are up to 8 cm tall by 8 mm wide and yellow-brown. Spore print is rust-brown. Widespread and common, Red-Gilled Cort fruits in wet spots in mixed woods, especially amongst moss. Also known as *Dermocybe semisanguinea.*

Cortinarius cinnamomeus

Cinnamon Cort

Caps are 2–4.5 cm across, bell-shaped to broadly convex, knobbed, dry, silky to minutely scaly, and cinnamon-yellow. Flesh is straw-coloured. Gills are attached, close, and yellowish to cinnamon. Stalks are up to 8 cm tall by 6 mm wide and yellowish. Ring is web-like and yellowish. Spore print is rusty-brown. Widespread and common, Cinnamon Cort fruits in wet spots under conifers. Also known as *Dermocybe cinnamomea.*

Cortinarius pholideus

Caps are 4–8 cm across, convex to bell-shaped or flat, with a broad knob, dry, and brown, covered with dark brown scales. Gills are attached, close, and violet, ageing brown. Stalks are up to 10 cm tall by 1.5 cm wide, coloured as the cap or paler, and scaly. Spore print is rust-brown. Widespread but not common, this species fruits under birch.

Cortinarius sanguineus

Blood-Red Cort

Caps are 2–4 cm across, bell-shaped to convex, knobbed, dry, silky to minutely scaly, and deep blood-red. Gills are attached, close, and coloured as the cap. Stalks are up to 8 cm tall by 8 mm wide and coloured as the cap. Ring is web-like and reddish. Spore print is rust-brown. Widespread but not common, Blood-Red Cort fruits under conifers. Also known as *Dermocybe sanguinea*.

MUSHROOMS WITH LIGHT SPORE PRINTS

In this group, the spore prints are white, cream-coloured or yellow. This group is easily the largest of the gill fungi and nearly two-thirds of the species illustrated in this book belong here. There are a host of interesting and spectacular genera represented and several of these contain more than 100 species. *Mycena* alone has over 200 species reported from North America. Thus, identification to genus in this group is not always easy and identification to species is possible only for species of distinction that can be recognized by a combination of size, colour, substrate, growth habit, shape, etc. I have selected for illustration those species that are most readily identified and/or most interesting, although, unfortunately, not always the most common. For example, Well's Amanita (*Amanita wellsii*, p. 235), with its flesh-coloured cap and powdery, lemon-yellow stalk, is one of the most distinctive mushrooms. I have seen it only once, however, in Upper Clements Wildlife Park in Nova Scotia.

Note on Edibility: Do not eat **any** mushroom in this group unless it is labelled edible. See p. 180 for further information on edibility.

Key to Genera of Light-Spored Mushrooms

FB = fruitbodies. See p. 181 for the details of a mushroom fruitbody.

1. Gills free (see p. 181) **2**
1. Gills attached, decurrent, or appear as shallow wrinkles **6**

2. Ring only, no cup or ridges at the base of stalk **3**
2. Cup or ridges present, with or without ring .... ***Amanita***

3. Caps and stalks slimy .... ***Limacella***
3. Caps and stalks dry **4**

4. Caps smooth .... ***Leucoagaricus***
4. Caps scaly **5**

5. Caps lemon-yellow .... ***Leucocoprinus***
5. Caps some other colour .... ***Lepiota, Macrolepiota***

6. Ring present on stalk **7**
6. Ring absent **9**

7. Gills running down stalk (decurrent), FB large, white .... ***Catathelasma***
7. Gills not decurrent **8**

8. Caps and stalks powdery .... ***Cystoderma***
8. Caps and stalks smooth or scaly .... ***Armillaria***

9. Gills thick and blunt or appearing as ridges, folds, or wrinkles **10**
9. Gills thin, well-formed **14**

10. FB blue-grey to purple-grey or black ... **11**
10. FB yellow to yellow-orange to orange or brown ... **12**

11. FB trumpet-shaped, surface nearly smooth, grey-black ... ***Craterellus***
11. FB purple or purple-brown, gills reduced to ridges ... ***Polyozellus, Gomphus***

12. FB bright orange, malformed, gills as ridges ... see ***Hypomyces***
12. FB yellow to orange-yellow or brown ... **13**

13. FB yellow/orange, with thick, decurrent gills ... ***Hygrophoropsis, Cantharellus***
13. Gills replaced with folds, ridges, or wrinkles ... ***Cantharellus, Gomphus***

14. Caps exuding milky fluid (latex) when bruised ... ***Lactarius***
14. Caps not exuding latex ... **15**

15. Gills very brittle, fracturing when thumbed ... ***Russula***
15. Gills not brittle ... **16**

16. Fruiting on conifer cones or old mushrooms ... **17**
16. Not on cones or old mushrooms ... **18**

17. Fruiting on cones ... ***Baeospora***
17. Fruiting on old mushrooms ... ***Collybia, Nyctalis***

18. Flesh thin but firm to tough, often reviving when re-wet ... **19**
18. Flesh thick or thin, but soft and shrivelling when dry, not reviving ... **27**

19. Stalks lateral ... **20**
19. Stalks central ... **22**

20. Gill edge smooth ... ***Marasmiellus, Panellus***
20. Gill edge ragged, irregular, or splitting when wet ... **21**

21. Gill edges splitting and recurving ... ***Schizophyllum***
21. Gill edges ragged or irregular ... ***Neolentinus, Lentinellus***

22. Gill edges ragged ... ***Neolentinus, Lentinus***
22. Gill edges even ... **23**

23. Fruits on wood ... **24**
23. Fruits on the ground ... **26**

24. Caps hairy to scaly ... ***Crinipellis***
24. Caps smooth ... **25**

25. Caps in dense clusters ... ***Gymnopus, Xeromphalina***
25. Caps scattered to cluster ... ***Marasmius, Micromphale***

26. Caps hairy to scaly ... ***Crinipellis***
26. Caps smooth to rough but not scaly ... ***Marasmius, Gymnopus***

27. Stalks lateral or absent, on wood ... **28**
27. Stalks central ... **32**

28. Caps tiny (< 1 cm wide in diameter) ... **29**
28. Caps larger ... **30**

29. Caps white ... ***Cheimonophyllum***
29. Caps grey ... ***Resupinatus***

30. Caps gelatinous ... ***Hohenbuehelia***
30. Caps not gelatinous ... **31**

31. Flesh of cap thin (1–3 mm) ... ***Marasmiellus, Phyllotus***
31. Flesh of cap thick (> 1 cm) ... ***Hipsizygus, Pleurotus***

32. Gills running down stalk (decurrent) ... **33**
32. Gills not decurrent ... **42**

33. Caps and/or gills waxy, fruits on the ground .... ***Hygrophorus, Hygrocybe***
33. Caps and gills not waxy, fruits on the ground or wood **34**

34. Caps 1–3 cm **35**
34. Caps larger **36**

35. FB fragile, breaking easily .... ***Rickenella, Mycena***
35. FB firm to tough, not readily broken .... ***Xeromphalina***

36. Fruits on ground **37**
36. Fruits on wood **39**

37. Caps yellow, orange, or yellow-orange .... ***Hygrophoropsis, Cantharellus***
37. Caps some other colour **38**

38. Caps grey to grey-black .... ***Pseudoclitocybe, Cantharellula***
38. Caps white to brown or blue-green .... ***Clitocybe***

39. Fruits in clusters **40**
39. Fruits solitary, scattered, or in small groups **41**

40. Caps bright yellow-orange .... ***Omphalotus***
40. Caps tan to pale brown .... ***Pseudoarmillaria***

41. Gills bright orange, thick .... ***Hygrophoropsis***
41. Gills yellowish .... ***Omphalina***

42. FB waxy .... ***Laccaria, Hygrophorus, Hygrocybe***
42. FB not waxy **43**

43. Caps tiny (< 3 cm in diameter), fragile, thin-fleshed **44**
43. Caps larger **46**

44. Fruits on wood **45**
44. Fruits on the ground .... ***Laccaria, Mycena***

45. FB in small clusters (3–15) .... ***Cyptotrama, Mycena***
45. FB in large clusters (> 20) .... ***Clitocybula***

46. Fruits on wood **47**
46. Fruits on ground **50**

47. Fruits in large numbers (cespitose) **48**
47. Fruits solitary or in small groups **49**

48. Caps red-brown, stalks dark and velvety below .... ***Flammulina***
48. Caps pallid to light brown, stalks white .... ***Clitocybula***

49. Gills yellow .... ***Tricholomopsis***
49. Gills white .... ***Megacollybia***

50. Stalks rooting .... ***Xerula***
50. Stalks not rooting **51**

51. Caps and/or gills pinkish to purplish .... ***Calocybe, Laccaria***
51. Caps some other colour **52**

52. Gills clearly notched at attachment point .... ***Tricholoma***
52. Gills not obviously notched **53**

53. Caps brown **54**
53. Caps whitish to purple or brown with purple cast .... ***Hygrophorus, Lepista***

54. Fruiting in clusters .... ***Lyophyllum***
54. Fruiting solitary or scattered .... ***Hygrophorus, Melanoleuca***

The Genus Amanita

Amanita spp. form a group of handsome and ecologically important fungi. Many are edible and highly regarded. Others, however, contain **deadly toxins** (see p. 319) and have caused the painful and premature deaths of many unfortunate people. Amatoxins, even in minute quantities, can cause permanent damage to vital organs, particularly the liver and kidneys. Therefore, because of the deadly nature of some species of *Amanita*, we do **not recommend** any as edible.

Amanita spp. are mostly large and often colourful. The classical features for recognition of an *Amanita* are **1.** white spore print **2.** free gills (not attached to the stalk) **3.** cup at the base and often patches or warts scattered over the cap **4.** and ring on the stem.

These features, however, are far from consistent for all species of *Amanita*. In some species (e.g., *A. vaginata*, p. 236), the ring is not present and in others it disappears early. The cup might be missing and represented merely by remnants or ridges at the base of the stalk, as in Fly Agaric (*A. muscaria*, p. 239) and some other species. The gills might be attached, rather than free, or become free only in age. Despite these inconsistencies, with a little experience it is fairly easy to recognize an *Amanita*.

Amanita spp. have a great ecological significance in forests, as suppliers of nutrients to both coniferous and hardwoods trees (see "Mycorrhizae–The Fungus Roots of Trees," p. 27). Fly Agaric is also a source of **hallucinogenic** compounds (muscimole and ibotenic acid) and is believed by some to be deeply involved in prehistoric rituals and the origins of religion.

Amanita porphyria

Amanita wellsii

Well's Amanita

Caps are 5–12 cm across, convex to flat, flesh-coloured to salmon-pink, with pale yellowish patches, and remnants of the inner veil adhering to the margin of the cap. Gills are free and white. Stalks are up to 12 cm tall by 2 cm wide, with a slightly bulbous base, lemon-yellow, and cottony-powdery. Ring disappears early. Cup is reduced to a yellow deposit around the basal bulb. Spore print is white. Widespread but rare and handsome and distinctive, Well's Amanita fruits on the ground in mixed woods.

Amanita ceciliae

Caps are 5–12 cm across, convex to flat or depressed, dark brown, fading in age, with large, white to greyish warts, and a striate margin. Gills are white, free, and close. Stalks are up to 12 cm tall by 12 mm wide and minutely scaly. Ring is absent. Cup is persistent and sac-like, enveloping the base. Spore print is white and non-amyloid. Widespread and not uncommon, this distinctive species fruits on the ground under conifers. *A. ceciliae* was previously known as *A. inaurata*. Edible, but **avoid** (see p. 234).

Amanita fulva

Caps are 4–10 cm across, conical to convex, becoming flat with a central knob, smooth, radially striate, and yellow-brown to orange-brown. Gills are free and white. Stalks are up to 12 cm tall by 12 mm wide, tapering towards the apex, and white. Ring is absent. Cup is prominent and whitish to tan. Spore print is white and non-amyloid. Common in low spots, this species often fruits near bogs and marshes. It is recognized by the conical, brown, smooth, striate cap. Edible, but **avoid** (see p. 234).

Amanita vaginata

Caps are 5–10 cm across, convex, becoming flat or knobbed, smooth, slippery when wet, with a striate margin, and variable in colour (grey to cinnamon or tan-brown). Flesh is white. Gills are free and white. Stalks are up to 15 cm tall by 1.5 cm wide and smooth or powdery. Ring is absent. Cup is sac-like, white to tan, irregularly roughened, and sometimes buried. Spore print is white and non-amyloid. Widespread and one of the most common of the *Amanita* spp., it is recognized by its lack of a ring and the grey or brown, striate cap. Edible, but **avoid** (see p. 234).

Amanita citrina

False Death Cap

Caps are 3–10 cm across, convex to flat, smooth or with scattered, buff patches, and pale greenish-yellow to yellowish-white. Gills are free, close, and white. Stalks are up to 12 cm tall by 1.5 cm wide, white, and smooth or slightly cottony towards the globose base. Ring is persistent and white to pale yellow. Cup is persistent and white to pale tan. Spore print is white and amyloid. Common and widespread, False Death Cap fruits on the ground in woods. The pale varieties are readily recognizable, but the deeper citron-yellow varieties can be confused with Death Cap (*A. phalloides*). **Poisonous**.

Amanita albocreata

Caps are 3–10 cm across, convex, becoming flat, white to pale yellow near the centre, and with small to large patches of the white to cream universal veil scattered over the cap. Stalks are up to 15 cm tall by 1.5 cm wide, with a large, globose basal bulb. Ring is absent. Cup is not obvious. Spore print is white and non-amyloid. Widespread but not common, this species fruits on the ground in woods. It is recognized by its colour and by its lack of a ring.

Amanita brunnescens

Cleft-Foot Amanita

Caps are 5–15 cm across, convex, becoming flat, dark brown, fading to pale brown or white, and covered with pale brown, cottony warts. Flesh is white, staining red-brown. Gills are free, white, and close. Stalks are up to 15 cm tall by 2 cm wide, with a cleft basal bulb, smooth, and white, staining red-brown. Ring is white, large, and persistent. Cup breaks into cottony fragments. Spore print is white and amyloid. Widespread and fairly common, Cleft-Foot Amanita fruits on the ground in woods. **Poisonous**.

Amanita flavoconia

Yellow Patches

Caps are 2–8 cm across, convex to flat, and bright yellow to orange-yellow, with scattered, yellow patches. Gills are free and white to yellowish. Stalks are up to 12 cm tall by 12 mm wide and white or pale yellow, with a bulbous base. Ring is persistent and cream to yellow. Cup disintegrates and remains in the soil or scattered on the stalk. Spore print is white and amyloid. Common and widespread, Yellow Patches fruits on the ground in woods. This species can be confused with Frost's Amanita (below).

Amanita frostiana

Frost's Amanita

Caps are 2–8 cm across, convex to flat, with a striate margin, and orange to orange-yellow, with yellow, floccose warts over the surface. Gills are white to yellowish and crowded. Stalks are up to 8 cm tall by 1 cm wide, white to yellowish, fibrillose to woolly, and with an ellipsoid basal bulb. Ring is drooping and yellowish. Cup remains as floccose, yellow patches at the base of the stalk. Spore print is white and non-amyloid. Frost's Amanita fruits on the ground in mixed woods. This species is not often reported because it is confused with Yellow Patches (above), which does not have a striate margin and has amyloid spores. **Poisonous.**

Amanita muscaria

Fly Agaric

Caps are 7–20 cm across, hemispherical to convex, becoming flat, with a striate margin, covered with white to buff, cottony patches, and yellow to orange-red or bright red. Gills are free, close and white to cream. Stalks are up to 15 cm tall by 2.5 cm wide, white or yellowish, and scaly near the base. Ring is large and white to cream. Cup is reduced to scaly rings at the base of the stalk. Spore print is white and non-amyloid. Widespread and very common, Fly Agaric fruits on the ground in woods. Yellow and orange-red varieties dominate in the Great Lakes region. Bright red varieties are absent around the Great Lakes and rare in eastern regions. **Poisonous.**

Amanita gemmata

Gem-Studded Amanita

Caps are 2–10 cm across, convex to flat or slightly depressed, sticky, slimy when wet, smooth or covered with white, cottony tufts, buff to dingy yellow-brown, and with a striate margin. Flesh is white. Gills are free, close, and white. Stalks are up to 12 cm tall by 1.5 cm wide, with a basal bulb, white, and smooth or nearly so. Ring is white and disappearing or attaching to the cap. Cup forms a basal collar. Spore print is white and non-amyloid. Not common, Gem-Studded Amanita fruits on the ground in mixed woods. **Poisonous.**

Amanita porphyria

Caps are 4–12 cm across, convex, becoming flat, smooth, sticky, and dark brownish-grey, with scattered, grey patches over the surface. Margin is non-striate. Gills are crowded, attached or free, and white to grey. Stalks are up to 12 cm tall by 2 cm wide, off-white, and striate or scaly. Ring is thin, grey, and striate above. Cup is reduced to grey remnants attached to the basal bulb. Spore print is white and amyloid. Widespread and not uncommon, this species fruits on the ground in mixed woods. It is recognized by its brown, non-striate, patchy cap. **Poisonous.**

Amanita rubescens

The Blusher

Caps are 5–15 cm across, convex, covered with cottony, grey or grey-pink scales, reddish or pale reddish-brown, fading to white, and with a reddish tinge. Flesh is white, staining reddish. Gills are free, close, and off-white, staining reddish. Stalks are up to 20 cm tall by 2 cm wide, with a bulbous base, white, and staining reddish. Ring is large, persistent, and white, with a pinkish tinge. Cup is inconspicuous or lacking, with fragments remaining in the soil. Spore print is white and amyloid. Widespread and not uncommon, The Blusher fruits under hardwoods and is recognized by its overall pinkish tinges. Edible, but **avoid** (see p. 234).

Amanita rhopalopus

Caps are 4.5–18 cm across, convex, becoming flat, dry, white to creamy, and covered with floccose, spiny to irregular warts. Margin is non-striate. Gills are crowded, nearly free, and white. Stalks are up to 18 cm tall by 2.5 cm wide and white, with a deeply rooting basal bulb. Ring is delicate and disappearing, sometimes attaching to the margin of the cap. Odour is unpleasant. Spore print is white and amyloid. Widespread but not common, this species fruits under conifers.

Photo: Brian Shelton.

Amanita virosa

Destroying Angel

Caps are 5–13 cm across, convex to flat, slippery when wet, smooth, and pure white. Gills are free, close, and white. Stalks are up to 15 cm tall by 2 cm wide, swollen at the base, smooth to silky or fibrillose, and white. Ring is large, persistent, white, and floppy. Cup is sac-like, encasing the base and part of the stalk. Spore print is white and amyloid. Widespread and common, this species fruits on the ground in woods. Destroying Angel is well-named as one of the prettiest and **deadliest** mushrooms. **Poisonous.**

Amanita bisporigera

In appearance this species is similar to Destroying Angel (above), but it has two-spored basidia (a microscopic characteristic). Many collections of *A. bisporigera* are more slender and delicate in appearance than the more robust Destroying Angel. Widespread and fairly common, this species also fruits on the ground in woods. **Deadly poisonous.**

Amanita farinosa

Caps are 2–8 cm across, convex to flat, striate, and covered with grey-brown, powdery-mealy remnants of the outer veil. Gills are free, white, and close. Stalks are up to 7 cm tall by 1 cm wide and whitish or with scattered grey-brown remnants of the ring, which disappears early. Spore print is white and non-amyloid. Widespread but not common, this species fruits on the ground in woods.

The Genus Lepiota and Similar Fungi

The distinctive features of *Lepiota* are **1.** white spore print **2.** free gills (not attached to the stalk) **3.** ring on the stalk **4.** outer layer of the cap breaks up into scales as the cap expands.

Some of the larger *Lepiota* spp. are now separated out into the genus *Macrolepiota*. A very common lawn mushroom, formerly called *Lepiota naucina*, is now called *Leucoagaricus naucina*; as the common name Smooth Parasol suggests, it differs because it lacks scales on the cap. A few of the small woodland *Lepiota* spp., such as *L. cristata* (p. 244) and Shaggy-Stalked Parasol (*L. clypeolaria*, p. 244), are easy to recognize, but a host of these small species are very difficult to tell apart using field characters. Some of these species are **poisonous**, so **none** of the *Lepiota* spp. are recommended as edible. *Macrolepiota procera* and *M. rhacodes* are edible and highly regarded, but be careful not to confuse them with the **green-spored** *Chlorophyllum molybdites* (not described in this book), which is poisonous. This situation is one where a spore print pays off!

Leucocoprinus luteus

Yellow Parasol

Caps are 3–6 cm, conical to bell-shaped, becoming convex in age, dry, lemon-yellow to sulphur-yellow, mealy or covered with reflexed scales, and radially striate. Gills are free and yellow. Stalks are up to 10 cm tall, slender, yellow, and with a swollen base. Ring is persistent or disappearing and yellow. Spore print is white. Widespread, Yellow Parasol fruits in flower pots and plant containers in greenhouses, malls, etc. Also known as *L. birnbaumii*.

Photo: Greg Thorn.

Lepiota acutaesquamosa

Sharp-Scaled Parasol

Caps are 5–12 cm across, convex to flat or knobbed, off-white to tan, dry, and covered with dark brown, pointed scales. Gills are free, crowded, white, and minutely toothed. Stalks are up to 12 cm tall by 12 mm wide, with a swollen base, white, and covered with cottony hairs. Ring is white, hanging loosely, and sometimes disappearing. Spore print is white. Widespread and not uncommon, Sharp-Scaled Parasol fruits on the ground or sometimes on wood in forests.

Lepiota rubrotincta

Red-Tinged Parasol

Caps are 2–8 cm across, convex to flat or knobbed, dry, at first a uniform pinkish-brown, and then breaking up towards the margin to form streaks of coral-pink to red-brown fibrils on a white base. Gills are white and free. Stalks are up to 10 cm tall by 8 mm wide, white, sometimes with a swollen base, and splitting readily. Ring is persistent and white to pinkish. Spore print is white. Widespread and not uncommon, Red-Tinged Parasol fruits on the ground in woods.

Limacella illinita

Caps are 2–7 cm across, conical to bell-shaped, with a low knob, slimy when wet, smooth, white, and sometimes yellowish near the centre. Gills are free, close, and white. Stalks are up to 8 cm tall by 5 mm wide, white, and slimy. Ring is hairy, becoming slimy and disappearing. Spore print is white. Widespread but not common, this species fruits on the ground in deciduous woods.

Lepiota clypeolaria

Shaggy-Stalked Parasol

Caps are 2–8 cm across, bell-shaped to convex, becoming flat, with a broad knob, dry, and brown at the centre, with brown, cottony scales on a cream base towards the outside. Gills are free, close, and white. Stalks are up to 9 cm tall, slender, white, and cottony or shaggy. Ring is white, cottony, and disappearing. Spore print is white. Widespread and common, Shaggy-Stalked Parasol fruits on the ground in woods, often in grassy spots. **Poisonous.**

Lepiota cristata

Caps are 1–5 cm across, bell-shaped, becoming flat, knobbed, dry, red-brown at the centre, and breaking up into scattered scales towards the margin. Gills are white and free. Stalks are up to 5 cm tall, slender, smooth to slightly hairy, and whitish. Ring is white and persistent or disappearing. Spore print is white. Widespread and common, this species fruits on the ground in woods or often in grass beside woodland trails. **Poisonous.**

Macrolepiota rachodes

Caps are 5–15 cm across, convex to flat or knobbed, dry, brown at first, and expanding and breaking up into coarse, brown scales on a white background. Flesh is white, bruising yellowish to brown. Gills are free and white, bruising brownish. Stalks are up to 20 cm tall by 2.5 cm wide, with a large basal bulb. Ring is white, large, and movable. Spore print is white. Widespread but not common, this species fruits in grass. Also known as *Lepiota rhacodes*. Edible.

Photo: Greg Thorn.

Macrolepiota procera

Parasol Mushroom

Caps are 8–25 cm across, conical to bell-shaped, becoming flat, with a broad knob, dry, and reddish-tan at the centre, with concentric rings of tan scales on a white base. Gills are free and crowded. Stalks are up to 40 cm tall by 1.5 cm wide, white, silky-striate, and covered with tan scales. Ring is persistent, large, flaring, and movable. Spore print is white. Widespread and not uncommon, this handsome species fruits in grass or in open woods. Also known as *Lepiota procera*. Edible. Photo: Brian Shelton.

Leucoagaricus naucina

Smooth Parasol

Caps are 4–12 cm across, convex, dry, smooth, and white, ageing buff. Gills are free, close, and white, ageing pinkish. Stalks are up to 10 cm tall by 13 mm wide, white, with a swollen base, and smooth. Ring is narrow and movable. Spore print is white. Widespread and common, Smooth Parasol fruits on lawns, sometimes in woods. This species is distinguished by its thin, movable ring and swollen base and the lack of a cup. Edible, but **not** recommended because it can be mistaken for the deadly Destroying Angel (*Amanita virosa*, p. 241).

Photo: Brian Shelton.

Armillaria mellea

Honey Mushroom

Caps are 3–10 cm across, convex, becoming flat, with a central knob, dry to sticky, covered with dark, scaly tufts, especially near the centre, and pale yellow-brown to rust-brown. Gills are attached to short-decurrent, close to well-spaced, and whitish, staining rusty. Stalks are up to 15 cm tall by 2 cm wide and coloured as the cap or paler. Ring is white, thick, and flaring. Spore print is white. Recognized by its colour, scaly cap, and flaring ring, this white-spored species, which is widespread and very common, fruits in clusters on wood or stumps in late fall. In recent years, Honey Mushroom complex has been broken up into a number of species that are sometimes difficult to tell apart in the field but **all** are edible.

Catathelasma ventricosa

Caps are 5–15 cm across, convex, becoming flat, white to off-white or greyish, thick, and fleshy. Gills are decurrent and coloured as the cap. Stalk is up to 10 cm tall by 5 cm wide and tapering to a narrow point, which is buried. Ring is double and flaring. Spore print is white. Widespread but not common, this species fruits on the ground under conifers. Edible.

Cystoderma amianthinum

Caps are 2–4 cm across, convex, dry, powdery to granular, and yellowish to orange-yellow, sometimes with radial wrinkles. Gills are attached, close, and white to cream. Stalks are up to 6 cm tall by 6 mm wide, powdery to granular, and coloured as the cap. Ring is thin and fragile, and the inner veil sometimes attaches to the margin of the cap. Spore print is white. Widespread and the most common of the *Cystoderma* spp., it fruits on the ground in woods.

Cystoderma granulosum

Caps are 2–5 cm across, convex to broadly convex, with a broad knob, dry, with a granular surface, radially wrinkled, and ochre to brick-red. Gills are attached, close, and white. Stalks are up to 5 cm tall by 8 mm wide, granular to scaly, and reddish-buff. Ring is narrow and disappears. Spore print is white. Widespread but not common, this species fruits on the ground in woods, often in moss.

Photo: Greg Thorn.

Cystoderma terrei

Caps are 2–8 cm across, convex, becoming flat, dry, mealy to granular, with numerous pointed scales, and orange-brown to rusty-brown. Gills are attached, becoming free, white to cream, and close. Stalks are up to 6 cm tall by 1 cm wide, mealy to granular, and cinnamon-brown. Ring is small and disappearing, and the inner veil sometimes attaches to the margin of the cap. Spore print is white. Widespread but not common, this species fruits on the ground under conifers. Also known as *C. cinnabarinum*.

Cantharellus and Related Mushrooms

The true chanterelles or "cantharelloid" fungi are characterized by **1.** a white spore print **2.** very thick gills or ridges. **3.** decurrent gills (gills run down the stalk) **4.** fruiting on the ground in woodlands. As is common in fungi, however, not all members of the group exhibit all of these diagnostic features.

This group contains some of the **highly prized edibles**, such as Chanterelle (*Cantharellus cibarius*, p. 250). The white-spored Horn of Plenty (*Craterellus cornucopioides*, below) is also placed here.

Pig's Ear (*Gomphus clavatus*, p. 249) and Woolly Chanterelle (*Gomphus floccosus*, p. 249) have decurrent wrinkles down the stalk in place of gills. Some consider Pig's Ear excellent to eat, but Woolly Chanterelle has caused **gastrointestinal** problems for some people. Both species of *Gomphus* have ochre spore prints, so they do not fit happily in the light-spored group. They are placed here because of similarities to the other cantharelloid fungi.

Craterellus cornucopioides

Horn of Plenty

Fruitbodies are 5–12 cm tall by 1–5 cm wide, trumpet-shaped, dry, smooth, with a wavy, inrolled margin, brown to smoky-grey or almost black, thin, and tough. Gills are absent or reduced to wrinkles. Stalks are not distinct, merging with the cap. Spore print is white. Widespread and fairly common, distinctive Horn of Plenty fruits on the ground in woods. Edible. Also see the similar *C. fallax* (p. 190).

Polyozellus multiplex

Caps are 1–5 cm across, funnel-shaped, dry, smooth but with an uneven surface and a wavy margin, purplish to black, and with a fragrant odour. Gills are decurrent, ridge-like, shallow, far apart, with cross veins forming a net-like arrangement, and grey. Stalks are smooth, 1–4 cm tall by up to 2.5 cm wide, often fused near the base, solid, and coloured as the cap. Spore print is white. Widespread but rare, this species fruits under conifers.

Photo: Greg Thorn.

Gomphus floccosus

Woolly Chanterelle

Fruitbodies are trumpet-shaped and up to 15 cm tall. Caps are 5–10 cm across, dry, with a flat or depressed top, yellow to orange with reddish tints, and covered with reddish to orange-red scales. Gills are decurrent, poorly developed, consisting of shallow wrinkles, and ochre to reddish-yellow. Stalks are smooth, short, running into the cap, and pale yellow. Spore print is ochre. Widespread and fairly common, this species fruits on the ground under conifers.

Gomphus clavatus

Pig's Ear

Fruitbodies are trumpet-shaped, up to 15 cm tall (including stalk) and 5–10 cm across at the top, dry, smooth to slightly scaly, with a flat or depressed top, sometimes with a wavy, lobed margin, and purple-brown. Gills are decurrent, poorly developed, consisting of shallow wrinkles, and pale purple-brown. Stalks taper to a narrow base, and are off-white to purplish. Spore print is ochre. Pig's Ear fruits on the ground under conifers. Edible.

Cantharellus cibarius

Chanterelle

Caps are 2–10 cm across, dry, smooth, convex, becoming flat and depressed, with a wavy margin, and yellow to orange-yellow. Gills are decurrent, far apart, thick, blunt, forked, and yellow. Stalks are up to 6 cm tall by 2 cm wide, smooth, and coloured as the cap or paler. Spore print is pale yellow. Widespread and fairly common, Chanterelle fruits on the ground in woods. It can be confused with *Hygrophoropsis aurantiaca* (p. 252). Edible.

Cantharellus cinnabarinus

Cinnabar Chanterelle

Caps are 1.5–5 cm across, convex, becoming depressed, with a wavy margin, dry, smooth, and cinnabar-red, fading in age. Gills are decurrent, far apart, blunt, forked, and coloured as the cap or tinged pink. Stalks are up to 4 cm tall by 6 mm wide, smooth, tough, and coloured as the cap or paler. Spore print is white. Widespread but not common and recognized by its thick gills and brilliant colour, Cinnabar Chanterelle fruits on the ground in woods. Edible.

Cantharellus tubaeformis

Caps are 1–5 cm across, convex to flat, becoming funnel-shaped, with a wavy margin, moist, olive to ochre or yellow-brown, and more or less covered with tiny, dark brown scales. Gills are decurrent, far apart, narrow, blunt, and ochre, becoming grey. Stalks are up to 6 cm tall by 6 mm wide, smooth, and yellow to ochre. Spore print is white. Easily recognized by its colour and thick, branching gills, this common and widespread species fruits on the ground in moist areas in woods. Edible.

Cantharellus lutescens

Smooth Chanterelle

Caps are 2–4 cm across, convex, umbilicate to funnel-shaped, dry, thin, smoky-brown to yellow-brown, becoming yellowish, and with an irregular, wavy margin. Gills are lacking, decurrent, and the underside is smooth or with shallow wrinkles. Stalks are up to 5 cm tall, merging with the cap, and yellow. Spore print is yellowish. Widespread and not uncommon, Smooth Chanterelle fruits on the ground, often in wet places.

Cantharellus minor

Caps are 1–4 cm across, convex, becoming flat and funnel-shaped, with a wavy margin, and bright orange to yellow-orange. Gills are decurrent and reduced to shallow ridges and wrinkles, with cross veins. Stalks are up to 5 cm tall, tapering from the cap to the base, coloured as the cap, and smooth. Widespread but not common, this species fruits in wet spots in deciduous woods.

Cantharellula umbonata

Grayling

Caps are 2–4 cm across, convex, becoming flat or depressed, often with a small, central knob, dry, smooth, and silvery-grey to smoky-grey. Flesh is white. Gills are decurrent, close, and white, staining reddish. Stalks are up to 8 cm tall by 7 mm wide and white to pale grey. Spore print is white. Widespread and not uncommon, Grayling fruits in moss and is recognized by its silvery cap, with a small knob.

Hygrophoropsis aurantiaca

False Chanterelle

Caps are 2–7 cm across, convex to flat, with an inrolled, paler margin, becoming depressed in age, dry, smooth to hairy, and orange to orange-brown. Flesh is buff to yellow-orange. Gills are decurrent, forked, and salmon to bright orange. Stalks are up to 8 cm tall by 13 mm wide, minutely hairy, and pale yellow to orange. Spore print is white. False Chanterelle fruits on rotten wood or on the ground in duff. Edible, but has caused **problems** for some people.

Rickenella fibula

Caps are tiny, 3–10 mm across, convex to flat or centrally depressed, dry, minutely hairy, faintly striate, and ochre-orange, fading in age. Gills are decurrent and white to yellowish. Stalks are up to 5 cm tall, slender, tough, and yellowish. Spore print is white. Widespread and common, this species fruits in mosses, often beside woodland trails. Also known as *Gerronema fibula*.

Chrysomphalina chrysophylla

Caps are 2–4.5 cm across, convex to flat, with a central depression, olive to yellow-brown, and minutely hairy to scaly. Gills are decurrent, well-spaced, and yellow to orange-yellow. Stalks are up to 4 cm tall by 3 mm wide, smooth, curved, and yellow. Spore print is yellow. Widespread but not common, this species fruits on conifer logs.

Photo: Brian Shelton.

Clitocybe clavipes

Club Foot

Caps are 2–8 cm across, convex to flat and depressed, often broadly knobbed, smooth, smoky-brown to grey-brown, and fragrant. Gills are decurrent, well-spaced, and white to cream. Stalks are up to 7 cm tall, enlarging downwards to a large, clavate base, and coloured as the cap. Spore print is white. Widespread and one of the most common of the *Clitocybe* spp., Club Foot fruits under conifers or in mixed woods.

Clitocybe gibba

Funnel Clitocybe

Caps are 4–8 cm across, convex, with an inrolled margin, becoming flat to depressed and finally funnel-shaped, dry, smooth, and tan, with a pinkish tinge, fading in age. Gills are decurrent, close, thin, and white. Stalks are up to 8 cm tall by 6 mm wide, with a cottony base, and coloured as the cap or paler. Spore print is white. Widespread and common, Funnel Clitocybe fruits on the ground in woods and is recognized by its funnel shape and pinkish-tan cap. Edible.

Clitocybe odora

Anise-Scented Clitocybe

Caps are 3–8 cm across, convex, becoming flat, with a broad knob, dry, smooth, grey-green to blue-green, and a fragrant odour. Gills are attached to decurrent, close, and white to buff. Stalks are up to 8 cm tall by 1 cm wide, white, and cottony at the base. Spore print is white. Widespread but not common, this distinctive blue-green species fruits on the ground in woods. Edible.

Clitocybe dealbata

Caps are 1–4 cm across, convex, with an inrolled margin, becoming flat, with a wavy margin, sometimes depressed, dry, smooth, thin, grey-brown when wet, and drying white. Gills are attached to decurrent, close, narrow, and whitish. Stalks are up to 3 cm tall by 5 mm wide, tough, powdery, and coloured as the cap. Spore print is white. Widespread but not common, this species fruits in grass. **Poisonous.**

Clitocybula abundans

Caps are 1–3 cm across, convex to flat, becoming depressed or umbilicate, dry, smooth, and pale grey-brown to whitish, with a darker centre. Gills are attached and white. Stalks are up to 6 cm tall, slender, coloured as the cap, and splitting readily. Spore print is white. Widespread but not common, this mushroom fruits in large clusters on conifer wood or on the ground from buried wood.

Clitocybula ocula

Caps are 2–4 cm across, convex to bell-shaped, becoming flat with a central depression, with the margin torn and split, and pale brownish to grey-brown. Gills are attached and whitish to greyish. Stalks are up to 6 cm tall by 6 mm wide. Spore print is white. Common and widespread, this species fruits on well-rotted logs in large numbers and is recognized by its pale grey-brown, lacerated cap.

Pseudoarmillariella ectypoides

Caps are 2–6 cm across, umbilicate to funnel-shaped, dry, thin, greyish to yellow-buff, radially striate with darker fibrils, and with tiny tufts of blackish scales at intervals. Gills are decurrent, narrow, well-spaced, and yellowish. Stalks are up to 6 cm tall, slender, coloured as the cap or paler, and hairy at the base. Spore print is white. Widespread but uncommon, this species fruits on decaying conifer logs. Also known as *Clitocybe ectypoides*.

Pseudoclitocybe cyathiformis

Caps are 2–6 cm across, flat and depressed, becoming funnel-shaped, dry, thin, greyish to blackish-brown, and radially striate with darker fibrils. Gills are decurrent, well-spaced, and white to yellowish. Stalks are up to 6 cm tall, slender, coloured as the cap, and hairy at the base. Spore print is white. Widespread but uncommon, this white-spored Clitocybe-like species fruits on decaying conifer logs. Also known as *Clitocybe cyathiformis*.

Photo: Greg Thorn.

Calocybe carnea

Pink Calocybe

Caps are 1.5–4 cm across, convex, becoming flat, and rose to violet-pink. Gills are attached, white, and close. Stalks are up to 4 cm tall by 7 mm wide. Widespread but not common, this species fruits in grass in clusters. The delicate rose-pink colour is distinctive amongst lawn fungi. Edible.

Collybia, Rhodocollybia and Gymnopus

In its traditional sense, *Collybia* includes all three of the above genera, and species of this complex are very common in both coniferous and deciduous woodlands. The taxonomy of the *Collybia* complex, however, has recently been reassessed and most of the traditional *Collybia* spp. are now in the genera *Rhodocollybia* and *Gymnopus*. Tuberous Collybia (*Collybia tuberosa*, p. 259) and *Collybia racemosa* (p. 259), both of which colonize and fruit on old mushrooms, are the species remaining in *Collybia* that are treated in this book. Species, formerly in *Collybia*, with pinkish-buff spore prints are now in *Rhodocollybia*. These include Spotted Collybia (*R. maculata*, p. 257) and Buttery Collybia (*R. butyracea*, p. 257).

Some "*Collybia*" spp. (e.g., *C. subnuda*) are firm to tough and can easily be confused with *Marasmius* complex. These species have now been placed in the genus *Gymnopus* and include Tufted Collybia (*G. confluens*, p. 258), Oak Collybia (*G. dryophilus*, below), *G. subnudus* (p. 259) and Clustered Collybia (*G. acervatus*, p. 258). To make the transition easier, I have continued to use the traditional common names for the species involved.

Gymnopus dryophilus

Oak Collybia

Caps are 2–6 cm across, convex to flat, smooth, dry, thin, often with a wavy margin, and tan to red-brown, fading in age. Gills are attached, narrow, crowded, and white. Stalks are up to 6 cm tall by 4 mm wide, smooth, often flattened, hollow, and tan to red-brown. Spore print is white. Widespread, Oak Collybia fruits on the ground under conifers or hardwoods and is one of the commonest woodland mushrooms. Also known as *Collybia dryophila*.

Rhodocollybia butyracea

Buttery Collybia

Caps are 2–7 cm across, convex to broadly convex, sometimes knobbed, slippery when moist, smooth, and reddish-brown, fading to tan. Gills are attached, close, and narrow, with a toothed edge. Stalks are up to 10 cm tall by 1 cm wide, smooth, striate, sometimes twisted, and light brown. Spore print is pinkish-buff. Widespread and common, this species fruits on the ground in conifer woods. Also known as *Collybia butyracea*. Edible.

Rhodocollybia maculata

Spotted Collybia

Caps are 5–15 cm across, convex, becoming flat, with a wavy margin, often broadly knobbed, dry, smooth, white, and with scattered, rusty stains. Gills are attached, narrow, crowded, and white. Stalks are up to 15 cm tall by 13 mm wide, brittle, and fibrous. Spore print is pale pinkish-buff. Widespread and not uncommon, Spotted Collybia fruits on the ground in woods. Also known as *Collybia maculata*. Edible.

Gymnopus confluens

Tufted Collybia

Caps are 2–5 cm across, convex to broadly convex, sometimes with a broad knob, dry, smooth to minutely scaly, buff to red-brown, drying pinkish-buff, tough, and reviving when wet (see p. 261). Gills are free, narrow, close, and white. Stalks are up to 10 cm tall by 5 mm wide, often compressed, tough, and red-brown underneath a coating of dense, white hairs. Spore print is white. Widespread but not common, Tufted Collybia fruits among fallen leaves and is recognized by its tufted habit and fuzzy stalks. Also known as *Collybia confluens.*

Gymnopus acervatus

Clustered Collybia

Caps are 2–5 cm across, convex to flat, dry, smooth, and red-brown, with a paler margin, ageing tan. Gills are close, narrow, and white or tinged pink. Stalks are up to 10 cm tall by 5 mm wide, smooth, hollow, and usually darker than the cap. Spore print is white. Widespread and common, Clustered Collybia fruits on the ground or on rotten wood. It is recognized by its clustered habit and tall stalks. Also known as *Collybia acervata*.

Gymnopus subnudus

Caps are 2–4 cm across, convex, becoming flat and depressed, with a small, central knob, striate, pale red-brown, and smooth. Gills are attached, well-spaced, and pallid to pale tan. Stalks are up to 5 cm tall by 2 mm wide, tough, roughened, and dark brown. Spore print is brown. Widespread and fairly common, this woodland species fruits on the ground in woody debris. Also known as *Collybia subnuda*.

Collybia tuberosa

Tuberous Collybia

Caps are 3–10 mm across, convex to flat, smooth, dry, and white to pinkish-tan. Gills are attached or decurrent, narrow, and white. Stalks are up to 2 cm long, slender, smooth to powdery below, and white to tan, arising from a polished, red-brown, hard body (sclerotium) several mm in diameter. Spore print is white. Widespread and common, Tuberous Collybia fruits on old, decayed mushrooms.

Collybia racemosa

Caps are 4–15 mm across, convex, becoming flat, grey to grey-brown, and paler near the margin. Gills are attached and grey to brownish. Stalks are up to 7 cm tall, slender, and bearing numerous short laterals, with swollen tips. Widespread but rare, this species fruits in clusters on old, decaying mushrooms.

Baeospora myosura

Spruce-Cone Mushroom

Caps are 0.5–2 cm across, convex, becoming flat, smooth, and tan, fading to whitish. Gills are attached, narrow, white to pale tan, and close. Stalks are up to 5 cm tall, slender, fuzzy, and whitish to tan. Spore print is white. Widespread and common, this delicate mushroom fruits on spruce cones during prolonged wet periods.

Nyctalis parasitica

Caps are 5–15 mm across, hemispherical to convex, white, becoming greyish to dingy brown, and radially striate to hairy. Gills are attached, far apart, narrow, thick, and off-white. Stalks are up to 3 cm tall, slender, powdery to silky, and coloured as the cap. Widespread but not common, this species fruits on decaying *Russula* spp. (pp. 310–15). Also known under its asexual name of *Asterophora parasitica*.

Nyctalis asterophora

Caps are 5–25 mm across, hemispherical to nearly globose, dry, and covered with a thick, powdery coating of brown dispersal spores (conidia). Gills are poorly developed and well-spaced. Stalks are up to 3 cm tall and slender. With its globose shape and powdery, brown spores on the cap, this mushroom has the aspect of a small puffball (*Lycoperdon* sp.). Widespread but rare, it fruits on old *Russula* caps. Also known as *Asterophora lycoperdoides*.

Photo: Richard Aaron.

The Genus Marasmius and Similar Mushrooms

Marasmius spp. and related fungi are best seen during or after heavy rains. They are called "**resurrection fungi.**" They dry, shrivel up and are more or less inconspicuous. They can remain in a dormant condition for extended periods (days to weeks). When re-wetted, however, the mushrooms flesh out and resume spore production. Most *Marasmius* spp. are fairly small and they can revive fairly quickly. The common Fairy Ring Fungus (*M. oreades*, p. 262) is a larger species of the genus. It grows in grassy places and is an excellent edible. *Crinipellis* (p. 266) looks like a hairy *Marasmius*. Some of the smaller *Collybia* spp. are easily confused with the larger *Marasmius* spp.

Micromphale foetidum

Caps are 1.5–4 cm across, convex, becoming flat, with a depressed centre, brown to red-brown, thin to membranous, and with dark, radial striations. Gills are attached to slightly decurrent, well-spaced, and with cross veins. Stalks are up to 4 cm tall by 2 mm wide, brown to dark brown, and velvety. Often with an unpleasant odour. Spore print is white. Widespread and not uncommon, this species fruits on hardwood twigs and branches. Also known as *Marasmius foetidus*.

Marasmius oreades

Fairy Ring Fungus

Caps are 2–5 cm across, bell-shaped to convex or flat to broadly knobbed, dry, smooth, honey-brown to tan, and fading in age. Margin is often striate. Flesh is thin, white, tough, and reviving when wet. Gills are nearly free, well-spaced, broad, and whitish to yellow-tinged. Stalks are up to 8 cm tall by 5 mm wide, tough, coloured as the cap or paler, and smooth or finely downy. Spore print is white. Widespread and common, this species fruits in grass. Edible.

Marasmius delectans

Caps are 5–35 mm across, convex, becoming nearly flat, dry, smooth, and yellowish-white. Gills are attached, becoming free in age, white, well-spaced, and with cross veins. Stalks are up to 6 cm tall by 2.5 mm wide, smooth, shiny, and yellowish-white above and dark brown near the base. Spore print is white. Widespread but not common, this species fruits in debris on the ground under hardwoods and often produces mycelial mats.

Marasmius cohaerens

Caps are 1.5–4.5 cm across, conical to broadly convex or bell-shaped, smooth, dry, and red-brown to yellow-brown, fading to pale tan. Gills are attached, off-white to pale brown, and well-spaced. Stalks are up to 8 cm tall by 3 mm wide, dry, and whitish above and dark brown to blackish-brown at the base. Spore print is white. Widespread and not uncommon, this species fruits early in the year in debris under hardwoods.

Marasmius siccus

Caps are 5–25 mm across, hemispherical to convex, dry, smooth, bright reddish-brown to orange-brown, and radially grooved. Flesh is white, thin, and revives when wet. Gills are free to attached, far apart, and white. Stalks are up to 7 cm tall, slender, tough, smooth, and blackish-brown. Spore print is white. Widespread and common, this species fruits on the ground in woods.

Marasmius pulcherripes

Caps are 4–12 mm across, convex to bell-shaped, smooth, dry, radially grooved, and bright pinkish-brown to red-brown. Flesh is thin, reviving when wet. Gills are attached, whitish, narrow, and well-separated. Stalks are up to 5 cm tall, wiry, and black. Spore print is white. Widespread but uncommon, this species fruits on the ground in mixed woods and is recognized amongst *Marasmius* spp. by its colour.

Marasmiellus nigripes

Caps are 5–20 mm across, hemispherical to convex, white, with a powdery surface, and delicate, radial wrinkles. Gills are attached to slightly decurrent, white, and well-spaced, with occasional cross veins. Stalks are up to 7 cm tall by 2 mm wide and dark bluish-grey, with a powdery surface. Spore print is white. Widespread but not common, this species fruits on the ground in mixed woods. Also known as *Tetrapyrgos nigripes*.

Marasmiellus candidus

Caps are 1–2.5 cm across, convex to flat, dry, white, with pinkish tints, radially grooved, thin, and reviving when wet. Gills are white, reddish-tinged, shallow, far apart, branching, and with cross veins. Stalks are up to 2.5 cm tall, slender, curved, and white to blackish-brown near the base. Spore print is white. Widespread and very common on the West Coast but rare in Ontario, eastern Canada and the adjacent U.S., this species fruits on dead sticks during wet periods. (For an additional illustration, see Fig. 10, p. 180.)

Marasmius capillaris

Caps are 2–6 mm across, convex, umbilicate, and whitish to pale tan, with deep grooves. Gills are attached, white, and distant. Stalks are up to 5 cm tall, wiry, black, tough, and shining. Spore print is white. This tiny but elegant species is widespread and not uncommon. It fruits during prolonged wet periods but shrivels quickly, so it is often missed.

Marasmius epiphyllus

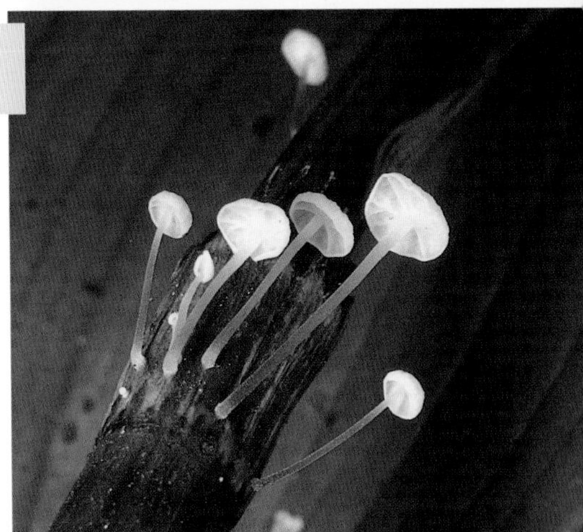

Caps are tiny, 3–8 mm across, convex, becoming flat, smooth, white, and with an irregular surface. Gills are white, narrow, and far apart. Stalks are up to 4 cm tall, slender, and brown. Spore print is white. Widespread and locally common, this tiny mushroom fruits on dead leaves or stalks of dead plants during prolonged wet periods.

Marasmius rotula

Horsehair Mushroom

Caps are 4–12 mm across, convex to hemispherical, umbilicate, dry, smooth, tough, thin, white, and radially grooved. Flesh is thin and reviving when wet. Gills are attached to a collar at the apex of the stalk, far apart, and white. Stalks are up to 5 cm tall, tough, very thin, dark, and shining. Spore print is white. Widespread and very common, this species fruits on decaying wood.

Marasmius androsaceus

Caps are 5–12 mm across, hemispherical to flat or depressed, dry, smooth, radially striate or grooved, and pinkish-brown to red-brown, often paler at the margin. Gills are attached, far apart, and flesh-coloured to pinkish-brown. Stalks are up to 8 cm long, narrow, tough, smooth, shining, and black. Spore print is white. Widespread and very common, this species fruits in troops on the ground in woods.

Crinipellis setipes

Caps are 5–12 mm across, convex to nearly flat, and whitish to tan but covered with darker brown scales and fibres. Gills are attached and white. Stalks are up to 3.5 cm tall, slender, blackish, and hairy. Spore print is white. Widespread and common, this species fruits on the ground or in debris in woodlands and is recognized by its fibre-streaked cap and hairy stalk.

Crinipellis zonata

Zonate Crinipellis

Caps are 1–2.5 cm across, convex, becoming flat, with a central knob, tan to dark brown, densely hairy to scaly, radially striate, and with weak, concentric zones in shades of brown. Gills are free, close, narrow, and white. Stalks are up to 5 cm tall, slender, and hairy. Spore print is white. Widespread and not uncommon, Zonate Crinipellis fruits on dead wood or woody debris.

Xeromphalina campanella

Caps are 0.5–2.5 cm across, convex to flat, depressed or umbilicate, dry, smooth, striate, and orange-yellow to yellow-brown, paler when dry. Gills are decurrent, narrow, far apart, and yellow. Stalks are up to 4 cm tall, slender, smooth, tough, yellow-brown to dark reddish-brown, and paler near the cap. Spore print is white to buff and amyloid. Widespread and common, this species fruits on decaying conifer wood in large clusters.

Xeromphalina cauticinalis

Caps are 5–25 mm across, convex, becoming flat, yellow-brown to orange-brown, smooth, centrally depressed to umbilicate, and radially striate. Gills are attached to slightly decurrent, well-spaced, with cross veins, and yellow. Stalks are up to 7 cm tall, slender, tough, and yellow-brown to blackish-brown. Spore print is white. Widespread and not uncommon, this species fruits on debris under conifers.

Xeromphalina tenuipes

Caps are 1–5 cm across, convex to flat, sometimes broadly knobbed, dry, velvety to mealy, and orange-brown, with an ochre or olive tinge. Gills are broadly attached to decurrent, white to pale yellow, and close. Stalks are up to 8 cm tall by 3–6 mm wide, coloured as the cap, and velvety-hairy. Spore print is white and amyloid. Widespread but not common, this species fruits on hardwood logs and stumps.

Photo: Greg Thorn.

The Genera *Hygrophorus, Hygrocybe* and *Camarophyllus*

Hygrophorus is a large genus accommodating approximately 250 species. Many of these mushrooms are extremely colourful and the colour range includes purple, green, scarlet, yellow, orange, rich chocolate-brown and snowy-white. They not only catch the eye with their colour, but many are extremely common and widespread in occurrence. This group, and the daintier *Mycena* spp. (pp. 282–88), are the most aesthetically pleasing of the mushrooms. A subgroup of *Hygrophorus* is split off by many mycologists into the genus *Hygrocybe* and a few other species into the genus *Camarophyllus*. *Hygrocybe* contains the smaller and more colourful species that often have a wax-like appearance. The larger, fleshier, more robust and usually less colourful species are contained in *Hygrophorus*. The differences between *Camarophyllus* and *Hygrophorus* are based on microscopic features.

Hygrophorus and *Hygrocybe* spp. grow on the ground in fields or forests. They don't fruit on wood. Many species of this group are edible, but have little flavour to recommend them. A **few** are believed **poisonous,** and because *Hygrophorus* spp. can be easily misidentified it is safer to say **not** recommended, even when you know it is edible. For these reasons, **none** of this group is recommended for eating.

Hygrocybe cantharellus

Chanterelle Waxcap

Caps are 0.5–3 cm across, convex to flat or depressed and umbilicate, with a scalloped margin, dry, flaky to minutely scaly, and bright orange-red, fading to orange or yellow. Gills are decurrent, well-spaced, broad, and usually paler than the cap. Stalks are up to 8 cm tall by 4 mm wide, smooth, and coloured as the cap or paler. Spore print is white. Chanterelle Waxcap fruits on the ground in moist woods.

Hygrocybe conica

Conic Waxcap

Caps are 2–5 cm across, conical, slippery when wet, smooth, bright red, fading to yellow or orange, often striate, bruising black, and splitting in age. Gills are free, well-spaced, and pale yellow, bruising black. Stalks are up to 9 cm tall by 6 mm wide, yellow to orange, staining black, splitting readily, striate, and often twisted. Spore print is white. Widespread and common, Conic Waxcap fruits on the ground in woods and is recognized by its shape and that it bruises black.

Hygrocybe acutoconica

Caps are 2–6 cm across, conical to bell-shaped, acutely knobbed, often with a flaring, upturned margin, very slimy, golden-yellow, and orange-tinged at the centre. Gills are attached and yellow. Stalks are up to 7 cm tall by 1 cm wide, coloured as the cap, fibre-streaked, and sometimes twisted. Spore print is white. Widespread but not common, this species fruits on the ground in mixed woods.

Hygrocybe punicea

Caps are 2–6 cm across, conical, becoming convex to nearly flat, slippery to slimy when wet, smooth, and bright red, becoming orange. Gills are attached, well-spaced, and yellow to orange-red. Stalks are up to 6 cm tall by 12 mm wide, becoming hollow, and red, fading to yellow, with a paler base. Spore print is white. Widespread and not uncommon, this species fruits on the ground in woods.

Hygrocybe miniata

Vermilion Waxcap

Caps are 1–3 cm across, convex to broadly convex, moist to dry, smooth, and red, becoming orange, then yellow. Gills are attached and coloured as the cap but paler. Stalks are up to 7 cm tall, slender, smooth, and coloured as the cap. Spore print is white. Widespread and fairly common, Vermilion Waxcap fruits on the ground in woods.

Hygrocybe flavescens

Yellow Waxcap

Caps are 2–6 cm across, convex to broadly convex, slippery when wet, shining when dry, smooth, and yellow to orange-yellow, fading to pale yellow. Margin is faintly striate. Flesh is pale yellow. Gills are attached to almost free, close or well-spaced, and yellow. Stalks are up to 8 cm tall by 1 cm wide, often flattened, and coloured as the cap but paler. Spore print is white. Widespread and not uncommon, Yellow Waxcap fruits on the ground in woods. Edible, but **not** recommended (see p. 268).

Hygrocybe coccinea

Scarlet Hood

Caps are 2–5 cm across, broadly conical to convex, becoming flat and knobbed, smooth, dry, and dull vermilion, fading to orange. Gills are attached, red to peach, becoming yellow at the edge, and well-spaced. Stalks are up to 4 cm tall by 6 mm wide and yellow, tinged with red. Spore print is white. Widespread and common, Scarlet Hood fruits under conifers.

Hygrocybe vitellina

Caps are 1–4 cm across, convex, becoming flat, with a down-turned, crimped margin, deeply umbilicate, bright yolk-yellow, and dry. Gills are coloured as the cap, decurrent, and far apart. Stalks are dry, up to 6 cm tall by 6 mm wide, coloured as the cap, and smooth. This species fruits under conifers. Seldom reported, it is thought by some to be the same as Nested Waxcap (*H. nitida*), but the latter is very slimy overall.

Hygrocybe ceracea

Caps are 5–25 mm across, convex to shallow-convex, with a depressed centre, orange-yellow to lemon-yellow, dry to slimy, thin, and striate. Gills are yellow, broadly attached to decurrent, and far apart. Stalks are up to 4 cm tall, slender, and coloured as the cap. Spore print is white. Rarely recorded, this small but attractive species fruits on the ground in mixed woods.

Hygrocybe virginea

Caps are 2–4 cm across, convex to flat and centrally depressed, white, and dry. Gills are decurrent, well-separated, and white. Stalks are up to 7 cm tall by 6 mm wide, white, and narrower towards the base. Widespread and not uncommon, this woodland species fruits in wet spots, often in moss, and is recognized by its dry, white cap.

Hygrophorus chrysodon

Golden Tooth Waxcap

Caps are 2–7 cm across, convex, becoming flat, with low knob, slippery when wet, shiny when dry, white, and dotted with minute, yellow granules, especially at the margin. Gills are decurrent, well-spaced, and white. Stalks are up to 7 cm tall by 1 cm wide and white, with yellow hairs near the apex. Spore print is white. Widespread but not common, Golden Tooth Waxcap fruits on the ground in woods and is recognized its yellow granules. Edible, but **not** recommended (see p. 268).

Hygrocybe nitida

Nested Waxcap

Caps are 1–3 cm across, convex, with an inrolled margin, becoming depressed and deeply umbilicate, slimy when wet, smooth, bright yellow, fading in age, and with a striate margin. Gills are decurrent, far apart, and yellow. Stalks are up to 7 cm tall by 4 mm wide, smooth, slimy, and coloured as the cap. Spore print is white. Widespread and locally common, Nested Waxcap fruits on the ground in moist woods.

Hygrocybe psittacina

Parrot Waxcap

Caps are 1–4 cm across, convex to bell-shaped, slimy, smooth, with a striate margin, and olive-green to pea-green, fading to yellow or ochre. Gills are attached, broad, well-spaced, and green to orange or yellow. Stalks are up to 7 cm tall, slender, smooth, sticky to moist, and green to yellow. Spore print is white. Widespread but not common, Parrot Waxcap fruits on the ground in woods.

Hygrophorus speciosus

Larch Waxcap

Caps are 2–8 cm across, conical to bell-shaped, slimy when wet, smooth, with an inrolled margin, red to orange-red, and yellow near the margin in age. Flesh is white. Gills are decurrent, far apart, broad, thick, and white to yellow. Stalks are up to 10 cm tall by 13 mm wide, sometimes flattened, and sticky. Spore print is white. Widespread but not common, Larch Waxcap fruits on wet ground under larch. Edible, but **not** recommended (see p. 268).

Hygrocybe laeta

Caps are 1.5–3 cm across, convex to flat, centrally depressed or umbilicate, with a striate, pale orange margin, slimy when wet, smooth, and carrot-orange, sometimes with an olivaceous tinge, and fading to apricot-yellow. Gills are decurrent and yellow to orange-yellow or greenish-yellow. Stalks are up to 10 cm tall by 2–5 mm wide, apricot-yellow to pinkish-buff, slimy, and smooth. Spore print is white. Widespread and not uncommon, this species fruits on the ground in moist, grassy spots in woods.

Hygrocybe subviolacea

Violet Waxcap

Caps are 2–5 cm across, broadly conical to bell-shaped, purple-grey to purple-brown, and sticky. Gills are decurrent, violet-grey, and well-separated. Stalks are up to 8 cm tall by 1 cm wide, coloured as the cap when young, and becoming whitish. Spore print is white. Violet Waxcap is widespread but not common and it fruits under conifers or in mixed woods. Also known as *H. lacmus*, *Hygrophorus subviolaceus*, and *Camarophyllus subviolaceus*.

Hygrophorus hypothejus

Caps are 3–7 cm across, convex to flat and knobbed, slimy when wet, smooth, olive-brown, paler towards the margin, and fading to ochre with reddish tints in age. Gills are decurrent, far apart, and yellowish. Stalks are up to 10 cm tall by 1 cm wide, coloured as the cap or paler. Ring is slimy and disappearing. Spore print is white. Widespread but not common, this species fruits on the ground under pines.

Photo: Greg Thorn.

Hygrophorus erubescens

Pink Waxcap

Caps are 4–10 cm across, convex to flat, knobbed, sticky, grey-white, with a pinkish tinge, to pink or red, and unevenly spotted with purplish-pink scales and hairs. Flesh is white to yellowish and bitter. Gills are attached, well-spaced, and spotted with red. Stalks are up to 7 cm tall by 1 cm wide and coloured as the cap. Widespread but not common, Pink Waxcap fruits under conifers.

Hygrophorus russula

Russula Waxcap

Caps are 5–10 cm across, hemispherical to convex, becoming flat or knobbed, slippery when wet, smooth or striate with darker hairs, fleshy, with an inrolled margin, and reddish to pink, darker at the centre. Gills are attached, becoming decurrent, close, and white to pink, staining purplish. Stalks are up to 7 cm tall by 3 cm wide, dry, and white, staining pink. Spore print is white. Widespread but not common, Russula Waxcap fruits under hardwood trees. Edible, but **not** recommended (see p. 268).

Hygrophorus pudorinus

Blushing Waxcap

Caps are 3–10 cm across, convex to bell-shaped, becoming flat, with a broad knob, slippery when wet, smooth, and pinkish to pinkish-tan to buff. Gills are attached to decurrent, thick, well-spaced, and white to cream, with a pink tinge. Stalks are up to 8 cm tall by 2 cm wide, dry, minutely roughened, and coloured as the cap. Spore print is white. Widespread and fairly common, Blushing Waxcap fruits on the ground in woods.

Hygrophorus fuligineus

Caps are 3–10 cm across, smooth, slimy when wet, and blackish-brown to dark olive-brown, paler at the margin. Gills are broadly attached to slightly decurrent, pallid, and far apart. Stalks are up to 10 cm tall by 1.5 cm wide, white, and slimy. Ring is slimy and disappears. Spore print is white. Widespread and not uncommon, this species fruits on the ground in woods.

Hygrophorus agathosomus

Almond-Scented Waxcap

Caps are 4–8 cm across, convex, becoming flat, sticky to slimy, smooth, and grey to brownish-grey. Bruised flesh has an almond odour. Gills are white to greyish, attached or short-decurrent, and well-spaced. Stalks are up to 8 cm tall by 2 cm wide, white to pale brown, and scaly and flaky near the top. Spore print is white. This species is widespread but not common in Ontario, eastern Canada and the adjacent U.S.

Hygrophorus camarophyllus

Caps are 2–9 cm across, convex, becoming flat, depressed to broadly knobbed, sticky to slimy, grey-brown to dark smoky-brown, paler towards the margin, and fibre-streaked. Gills are decurrent, white to greyish, and well-spaced. Stalks are up to 10 cm tall by 2 cm wide and whitish or streaked with the cap colour. Spore print is white. Widespread but not common, this species fruits under conifers.

Hygrophorus bakerensis

Caps are 5–12 cm across, hemispherical, cinnamon to brown, darker at the centre, becoming flat, and slimy when wet, with brownish scales below the slime. Bruised flesh has the odour of almonds. Gills are attached, becoming decurrent, and white to cream. Stalks are up to 12 cm tall by 2.5 cm wide and white to cream. This species fruits under conifers, it has been recorded, in Ontario, eastern Canada and the adjacent U.S., but it is not common. Edible, but **not** recommended (see p. 268).

Camarophyllus pratensis

Meadow Waxcap

Caps are 2–7 cm across, convex, sometimes broadly knobbed or depressed, dry, smooth, and pale red-brown to orange-brown, fading to tan. Gills are decurrent, far apart, and flesh-coloured. Stalks are up to 7 cm tall by 2 cm wide, dry, and whitish or coloured as the cap. Spore print is white. Widespread but not common, Meadow Waxcap fruits on the ground in woods. Also known as *Hygrophorus pratensis*. Edible, but **not** recommended (see p. 268).

Hygrocybe marginata

Orange-Gill Waxcap

Caps are 2–5 cm across, conical to bell-shaped, becoming convex to flat and broadly knobbed, smooth, and orange, fading to pale yellow. Gills are attached and bright orange. Stalks are up to 8 cm tall by 6 mm wide, smooth, sometimes flattened, and coloured as the cap. Spore print is white. Widespread but not common, Orange-Gill Waxcap fruits on the ground in woods. Also known as *Hygrophorus marginatus* and *Humidicutis marginatus*.

The Genus Laccaria

Laccaria spp. is a common mycorrhizal fungus associated with a number of trees. While there are only a few common species, they occur frequently in a wide variety of habitat types. *Laccaria* spp. are small, with white spores and attached gills. The gills are flesh-coloured to purple, tend to be fairly well-spaced, and are thickish and waxy. Common Laccaria (*L. laccata*, p. 279) has a distinctive violet-red colour that distinguishes it from all other mushrooms. Other species of the genus are similar in colour, except for Purple Laccaria (*L. amethystea*, p. 279), which is very variable but can be violet to rich purple.

Laccaria tortilis

Caps are tiny, 3–15 mm across, convex to flat or funnel-shaped, with a wavy margin, striate, smooth, and pinkish-ochre, paler towards the margin. Gills are violet-pink and well-spaced. Stalks are short, mostly less than 10 mm tall, slender, and coloured as the cap. Spore print is white. Widespread but not common, this species fruits on bare ground in wet spots in woods. For those enthusiasts with microscopes, the spores are large (up to 14 µm diam) and spiny.

Laccaria trullisata

Sand-Loving Laccaria

Caps are 2–5 cm across, convex to broadly convex and depressed or umbilicate, dry, fibrous to minutely scaly, especially near the margin, and red-brown to dull brown. Gills are attached, thick, well-separated, and purplish, ageing red-brown. Stalks are up to 8 cm tall by 2 cm wide, fibrous-striate, and red-brown. Spore print is white. Not common, Sand-Loving Laccaria fruits in sand or sandy soil.

Photo: Greg Thorn.

Laccaria laccata

Common Laccaria

Caps are 2–6 cm across, convex to flat and umbilicate or depressed, dry, smooth to minutely scaly, and bright red-brown to brick-red when wet, paler when dry. Gills are attached to slightly decurrent, well-spaced, thick, broad, and white to flesh-coloured or pinkish-brown. Stalks are up to 8 cm tall by 8 mm wide, dry, smooth, fibrous-striate, and coloured as the cap. Spore print is white. Widespread and very common, Common Laccaria fruits on the ground in woods. Edible.

Laccaria amethystea

Purple Laccaria

Similar in general habit to Common Laccaria (above), this species differs in its violet to purple colour. Widespread but not common, other varieties of this species are dull purple, and many have a whitish cast. Edible.

Laccaria ochropurpurea

Caps are 4–15 cm across, convex, becoming flat to depressed in the centre, smooth, dry, and dirty white to purplish-brown. Gills are attached, well-spaced, thick, and violet to purple. Stalks are up to 15 cm tall by 2 cm wide, coloured as the cap, and smooth to slightly scaly. Spore print is violaceous-white. Widespread but not common, this species fruits under hardwoods. Edible.

Lepista nuda

Blewit

Caps are 5–12 cm across, convex, becoming flat, then depressed, with a wavy outline, and with a pleasant odour. Colour varies from bright purple to faded violet or brownish. Gills are violet to purple. Stalks are up to 9 cm tall by 2.5 cm wide, coloured as the cap, and fibrous-streaked. Spore print is white. Widespread and common, Blewit fruits on the ground in woods. Edible.

Lepista irina

Caps are 5–12 cm across, convex to flat, with a wavy margin, smooth, dry, pinkish-white, becoming tan, and with a pleasant odour. Stalks are up to 10 cm tall by 2 cm wide, dry, white, becoming tan, and fibrous-striate. Gills are attached and flesh-coloured to tan. Spore print is pinkish-buff. Widespread and common, this species forms large fairy rings in woods. Edible.

Lyophyllum decastes

Fried Chicken Mushroom

Caps are 5–15 cm across, convex to flat, then depressed, with a wavy outline, and grey-brown to smoky-brown or reddish-brown. Gills are attached, white, and crowded. Stalks are up to 7 cm tall by 2.5 cm wide, white, and smooth. Spore print is white. Widespread and common, Fried Chicken Mushroom fruits in tufts on the ground in woods, often at the edge of trails or woodland parking lots. Edible.

Melanoleuca melaleuca

Caps are 3–7 cm across, convex to almost flat, with a low knob, moist, smooth, and blackish-brown to smoky-brown when wet, drying to tan. Flesh is white. Gills are attached, close, and white. Stalks are up to 8 cm tall by 1 cm wide, white, and streaked with dark fibrils. Spore print is white and amyloid. Common and widespread, this species fruits in grassy places in woods. It is recognized by the erect stalk, with a dark brown, flat cap and central knob. Edible.

Mycena and Related Mushrooms

Mycenas are amongst the daintiest and most attractive of the mushrooms. For the most part, they are tiny (measured in millimetres) or small (1–3 cm). They attract the eye, however, because they can fruit in spectacular and colourful clusters from rotting wood or extend in troops of hundreds scattered across the forest floor. This genus is large, with hundreds of species described and not all obviously belong to *Mycena*. Again, with such a large number of species, it becomes difficult to tell them apart and the majority of *Mycenas* will remain nameless to most of us. There are, nevertheless, a number of species that are common and reasonably distinctive. They often grow on wood in clumps of 3 to 20. If you find a small mushroom on wood in larger numbers, check *Clitocybula* (p. 254) or *Collybia* (p. 259). *Mycenas* are, for the most part, too small and fragile to be of value as edibles but are very **photogenic**.

Generally, Mycena spp. have the following features: **1.** white-spored **2.** have attached gills **3.** are fragile **4.** have conical or hemispherical caps **5.** are tiny to small.

Mycena subcaerulea

Caps are 5–15 mm across, conical to bell-shaped, becoming flat, sticky, smooth, striate, and blue to greenish-blue, ageing grey-brown. Gills are attached and white to greyish. Stalks are up to 8 cm tall, slender, powdery to minutely hairy, and coloured as the cap. Spore print is white. Widespread but not common, this species fruits on decaying hardwood.

Photo: Greg Thorn.

Mycena amabillisima

Caps are 3–20 mm across, broadly conical to bell-shaped, sometimes convex, with a knob, smooth, slippery when wet, and bright pinkish-red, fading to cream or white in age. Gills are attached, far apart, and white or tinged red. Stalks are up to 5 cm tall, slender, and powdery becoming smooth. Spore print is white. Widespread and locally common, this species fruits under conifers in wet spots, often in *Sphagnum* moss.

Mycena adonis

Caps are 5–20 mm across, bell-shaped, salmon-pink to rose-red, and striate. Gills are well-spaced, whitish, and tinged pink. Stalks are up to 3 cm tall, slender, and white to pinkish near the cap. Spore print is white. Widespread and locally common, this attractive *Mycena* fruits in wet spots, often in moss, under conifers.

Mycena rorida

Caps are 5–10 mm across, convex to broadly convex, becoming flat, with a central depression, pale brownish, fading to tan and finally yellowish-white, striate, with a scalloped margin, dry, and minutely roughened. Gills are attached, well-spaced, and whitish. Stalks are up to 5 cm tall, slender, whitish, and covered with a thick, slimy sheath. Widespread and not uncommon, this species fruits on conifer debris. Also see p. 22.

Mycena viscosa

Caps are 8–15 mm across, convex, slimy, radially striate, yellowish-grey to greenish-grey or brown to olive, and with an unpleasant taste. Gills are attached and yellowish-white to greenish-grey. Stalks are up to 7 cm tall by 3 mm wide, lemon-yellow to greenish-yellow, and smooth. Widespread and fairly common, this species fruits on the ground in mixed woods.

Mycena galericulata

Caps are 2–4 cm across, conical to bell-shaped, becoming flat or knobbed, smooth, striate, dark brown in the centre, paler towards the margin, and fading to tan when dry or with age. Gills are attached, white, tinged with pink, and well-spaced. Stalks are up to 10 cm tall, slender, smooth, tough, and grey to red-brown. Spore print is white. Widespread and common, this species fruits on rotten wood in dense clusters.

Mycena luteopallens

Caps are 8–15 mm across, conical to bell-shaped, becoming flat with a knob, striate when wet, and orange to yellow. Gills are attached and yellowish or tinged pink. Stalks are up to 9 cm tall, slender, and coloured as the cap. Spore print is white. This species fruits on old hickory nuts and walnuts and follows the distribution of these trees.

Mycena osmundicola

Caps are 3–6 mm across, convex, white, dry, powdery, and faintly striate. Gills are well-spaced, attached to nearly free, and white. Stalks are up to 3 cm tall, slender, white, and hairy, becoming smooth. This attractive species is not uncommon and fruits on the ground under pines.

Mycena pura

Caps are 2–4 cm across, convex to flat or broadly knobbed, moist to dry, smooth, with a striate margin, and pink to violet or purplish-brown. Gills are attached, white to lilac, and well-spaced. Stalks are up to 8 cm tall by 3–6 mm wide, smooth, and coloured as the cap or paler. Spore print is white. Widespread and common, this species fruits in small groups on the ground in woods.

Mycena haematopus

Caps are 1–3 cm across, bell-shaped, wine-red, smooth, striate, and with a toothed or scalloped margin. Gills are attached and whitish, staining reddish. Stalks are up to 8 cm tall, slender, fragile, smooth with a hairy base, coloured as the cap, and exuding reddish juice when broken. Spore print is white. Widespread and common, this species fruits in dense clusters on rotten wood.

Mycena rosella

Caps are 0.5–1.5 cm across, hemispherical to convex, rose-pink to greyish-pink, smooth, and radially striate. Gills are attached and pink with a darker edge. Stalks are up to 6 cm tall and slender. Spore print is white. Widespread and not uncommon, this species fruits under conifers.

Mycena stylobates

Caps are 3–15 mm across, conical to convex, becoming flat, with a recurved margin, whitish to pale grey, and striate. Gills are attached, becoming free in age, and whitish to pale grey. Stalks are up to 6 cm tall, slender, coloured as the cap, scaly and flaky, and attached to the substrate by a well-marked basal disc. Widespread but not common, this species fruits on plant debris.

Mycena epipterygia

Caps are 1–2.5 cm across, conical to bell-shaped, smooth, shiny, yellow to grey or brownish, and with a striate margin. Gills are attached and white, ageing brownish. Stalks are up to 7 cm tall, translucent yellow, and slender. Spore print is white. Widespread and common, this species fruits on conifer wood or under conifers. *M. viscosa* (p. 284) also has a yellow stalk, but is much more robust.

Mycena leaiana

Caps are 2–4 cm across, convex, slippery when wet, smooth, with a striate margin, and bright orange, fading to almost white in age. Gills are attached, yellow, with orange-red edges, and close. Stalks are up to 8 cm tall, slimy, slender, smooth, and coloured as the cap. Spore print is white. Common and widespread, this species fruits on wood, usually in clusters.

Mycena polygramma

Caps are 2–3 cm across, conical, becoming bell-shaped, knobbed, brown to grey-brown, smooth, and radially striate. Gills are white to greyish. Stalks are up to 10 cm tall by 5 mm wide, frequently grooved and slightly twisted, scaly and flaky, and white on a grey-blue background. Widespread but not common, this species fruits in clusters on well-rotted logs and stumps.

Mycena strobilinoides

Caps are 1–2 cm across, acutely or broadly conical, becoming bell-shaped, striate at the margin, slippery when wet, smooth, and bright red, fading to yellow. Gills are attached to slightly decurrent, yellow to salmon, and with a bright red margin. Stalks are up to 4 cm tall, slender, orange to apricot, and powdery, becoming smooth. Spore print is white. Widespread and locally common, this species fruits under pine.

Photo: Greg Thorn.

Flammulina velutipes

Velvet Stalk

Caps are 2–6 cm across, convex to broadly convex, slimy when wet, smooth, and yellow-brown to red-brown and darker at the centre. Gills are attached, well-spaced, and cream to yellow. Stalks are up to 7 cm tall by 6 mm wide, tough, velvety, and turning dark brown to almost black. Spore print is white. Widespread and fairly common, Velvet Stalk is confirmed by its clustered habit on wood, its velvety stalk, and that it fruits in cold weather. Edible.

Cyptotrama asprata

Caps are up to 2.5 cm across, convex to broadly convex and depressed, dry, minutely warty, becoming pitted and wrinkled in age, and yellow to orange-yellow, darker at the centre. Gills are broadly attached to decurrent, far apart, and white to yellowish. Stalks are 6 cm tall, slender, with a granular surface, and coloured as the cap or paler. Spore print is white. Widespread and common, this species fruits on decorticated wood.

Pleurotus and Similar Mushrooms

The "pleurotoid" mushrooms are characterized by features associated with the Oyster Mushroom (*Pleurotus ostreatus*, p. 291). Pleurotus-like features include **1.** a white spore print **2.** the stalk is lacking, off-centre or lateral **3.** they fruit shelving on wood.

Pleurotus is usually a fairly robust mushroom, with thick flesh. Angel's Wings (*Phyllotus porrigens*, p. 292) is similar to *Pleurotus*, but is so thin-fleshed that, in strong light, it is almost transparent. *Hohenbuehelia* is similar to *Pleurotus* or *Phyllotus* but has a gelatinous layer just below the cuticle. *Panus*, *Lentinus* and *Lentinellus* are also white-spored, shelving mushrooms but are tougher and have serrated gill edges. *Cheimonophyllum* is a tiny, white, pleurotoid mushroom measured in millimetres. It resembles *Crepidotus*. *Crepidotus* looks like it might belong with this group, but although white when young it has brown spores and the gills turn brown at maturity.

Hohenbuehelia angustata

Caps are 2–5 cm across, fan-shaped to spatula-shaped, sometimes inrolled and funnel-shaped, shelving, whitish to pinkish-buff, and with a gelatinous upper layer. Gills are narrow and whitish, ageing buff. Stalks are absent. Spore print is white. Widespread but not common, this species fruits on hardwood stumps and logs. Distinguished from *Crepidotus* (pp. 220–21) by the white spore print and from *Pleurotus* by the gelatinous layer. Edible.

Pleurotus ostreatus

Oyster Mushroom

Caps are 5–20 cm across, convex to flat, shelving, shell-shaped to semi-circular, moist, smooth, and whitish to grey-brown or violet-brown. Flesh is thick and white. Gills are decurrent, close to well-spaced, and white. Stalks are lateral, short, stout, white, hairy near the base, and often lacking. Spore print is white. Widespread and common, Oyster Mushroom fruits on dead logs or stumps and sometimes standing trees. Edible.

Pleurotus dryinus

Caps are 5–15 cm across, convex to flat, with a depressed centre, inrolled margin, and dry, suede-like surface, and white to cream, staining yellow. Gills are decurrent, coloured as the cap, and well-separated. Stalks are up to 10 cm tall by 3 cm wide, white to cream, and off-centre. Ring disappears, with remnants forming a border around the margin of the cap. Spore print is white. Widespread but not common, this distinctive species fruits on hardwood stumps and logs. Edible.

Phyllotus porrigens

Angel's Wings

Caps are 1–8 cm across, fan-shaped, with an inrolled margin, expanding to almost flat, shelving, pure white, dry, and smooth to hairy near the attachment. Flesh is thin and white. Gills are narrow, close, and white. Stalks are absent. Spore print is white. Widespread and not uncommon, Angel's Wings fruits on conifer wood. It is similar to Oyster Mushroom (p. 291), but is very thin-fleshed. Also known as *Pleurocybella porrigens*. Edible.

Hypsizygus tessulatus

Caps are 5–12 cm across, convex, with an inrolled margin, becoming flat, moist to dry, smooth, and white to buff, with a pinkish tinge. Flesh is white and thick. Gills are attached, broad, and white. Stalks are up to 7 cm tall by 2 cm wide, lateral, stout, solid, and smooth or hairy. Spore print is white. Widespread but not common, this species fruits on the wood of deciduous trees. Also known as *Pleurotus ulmarius*. Edible.

Lentinellus ursinus

Caps are bracket-form, 3–10 cm at the base, up to 5 cm wide, more or less flat, woolly near the attachment, pallid to tan, and sometimes becoming dark brown to blackish near the centre. Flesh is thin, white to buff, and tastes bitter. Gills are close to well-spaced, with ragged edges, and pallid to pinkish-buff. Stalks are absent. Spore print is white. Widespread and not uncommon, this species fruits on old logs or stumps.

Panellus serotinus

Late Fall Oyster

Caps are 2–10 cm across, semicircular to kidney-shaped, slippery when wet, smooth, yellow-green to olive, and with an inrolled, striate margin. Gills are thin, close, and white to pale tan-yellow. Stalks are short, broad, and laterally attached. Spore print is white and amyloid. Widespread and common, Late Fall Oyster fruits on hardwoods. Edible.

Lentinellus cochleatus

Caps are 3–6 cm across, deeply funnel-shaped, with an inrolled margin, and reddish-brown to pallid grey-brown. Gills are pallid, decurrent, and well-spaced. Stalks taper sharply, and are fused at their bases. They are coloured as the gills, but are darker towards the base. Widespread but not common, this species fruits in tight clumps on hardwood stumps. Edible.

Photo: Greg Thorn.

Neolentinus adhaerens

Caps are 2–7 cm across, convex, becoming broadly knobbed, circular to kidney-shaped, ochre to red-brown at the centre, and paler towards the inrolled margin. Gills are white to yellowish, well-spaced, and uneven, with an irregular edge. Stalks are up to 5 cm tall by 1 cm wide, central to off-centre, and coloured as the cap. Widespread but rare in Ontario, eastern Canada and the adjacent U.S., this species fruits on dead conifer wood. Also known as *Lentinus adhaerens*.

Omphalotus olearius

Jack O'Lantern

Caps are 5–15 cm across, convex, becoming flat or depressed, often with a broad knob, dry, smooth, streaked with darker fibrils, orange-yellow, and with a wavy margin. Gills are decurrent, close, narrow, and orange-yellow. Stalks are up to 20 cm tall by 2 cm wide, often twisted, smooth, and whitish to orange-pink. Spore print is cream. Widespread and locally common, Jack O'Lantern fruits on stumps or buried wood in dense clusters. This species glows in the dark, hence the common name. Also known as *O. illudens*. **Deadly poisonous.**

Photo: Tom Hsiang.

Schizophyllum commune

Split Gill

Caps are 6–25 mm across, fan-shaped, dry, white to grey, densely hairy, thin, tough, with an inrolled margin, shrivelling when dry, and reviving when wet. Gills are few, thick, appearing greyish and hairy when dry, and smooth and creamy white when wet (gills split down the middle when wetted). Stalks are lacking. Spore print is cream to yellow. Widespread and common, Split Gill fruits on dead wood.

Panellus stipticus

Luminescent Panellus

Caps are 1–5 cm across, shell-shaped to kidney-shaped, with a lobed margin, dry, tough, tan to brown, hairy to mealy, becoming smooth, and with a bitter taste. Gills are thin, close, and ochre to cinnamon. There is a clear line of demarcation between the gills and stalk. Stalks are short, lateral, flattened, and pallid. Spore print is white and amyloid. Widespread and common, Luminescent Panellus fruits on dead hardwood in overlapping clusters. It glows in the dark.

Plicaturopsis crispa

Caps are tiny, 5–20 mm across, shelving, tough, white and hairy when young, becoming tan-brown to red-brown in age, and curling when dry, reviving when wet. Gills are off-white to blue-grey, shallow, and blunt. Stalks are lacking. Spore print is white. Widespread and common, this species fruits on dead branches of hardwoods and is seen most often during or after rainy periods when it revives.

Resupinatus applicatus

Caps are tiny, up to 6 mm across, cup- to saucer-shaped, dry, minutely hairy, tan to dark grey, becoming black, and with an inrolled margin. Flesh is thin and gelatinous. Gills are tan to grey, far apart, thick, and blunt. Stalks are absent or inconspicuous and powdery to cottony at the base. Spore print is white. Widespread but not common, this species fruits on the underside of logs or branches.

Photo: Greg Thorn.

Cheimonophyllum candidissimum

Caps are tiny, 4–10 mm across, semi-circular to spatula-shaped, chalk-white, and delicate. Stalks are short and lateral. Gills are white, narrow, and well-spaced. Spore print is white. Widespread but not common, this species fruits on dead hardwood branches. It looks like a small *Crepidotus* (pp. 220–21), but the latter has brown spores and its gills turn brown at maturity.

Omphalina ericetorum

Caps are 5–25 mm across, convex to flat, dry to moist, radially striate, and yellow-brown to pale yellow or whitish. Gills are decurrent, white, broad, and far apart. Stalks are up to 2.5 cm tall, slender, smooth, and pale buff to pale yellow. Spore print is white. Widespread but not common, this species fruits on wood or on the ground in moss.

Tricholoma and Similar Mushrooms

Tricholoma is a fairly large genus with about 100 known species. For the most part, they are important biologically in mycorrhizal associations with forest trees, particularly conifers. They tend to fruit late in fall, often after the first frost. In certain cases some of the grey Tricholomas, such as *T. myomyces* (p. 299), come up in such abundance in pine woods that they could be collected by the bushel basket. Unfortunately, these grey *Tricholoma* spp. are difficult to tell apart and some are known to be **poisonous**. Therefore, **none** are recommended for eating.

The common characteristics of *Tricholoma* spp. are **1.** fleshy and medium-sized to large. **2.** white-spored **3.** gills attached and notched **4.** fruit on the ground, never on wood **5.** fruit late in the season **6.** ring is absent.

Tricholomopsis is like a *Tricholoma* but grows on wood. They are reported as edible but of undistinguished flavour. *Megacollybia* also grows on wood and was previously in *Tricholomopsis*. Although also reported as edible, *Megacollybia* has caused **gastrointestinal** upsets in some people.

Tricholoma flavovirens

Yellow-Green Tricholoma

Caps are 5–10 cm across, convex, becoming flat, sometimes broadly knobbed, slippery when wet, smooth to minutely scaly at the centre, and pale yellow to bright yellow, with brownish or olivaceous tints, especially near the margin. Flesh is white to yellowish. Gills are free, close, and sulphur-yellow. Stalks are up to 8 cm tall by 2 cm wide, white to pale yellow, and more or less smooth. Spore print is white. Widespread and common, Yellow-Green Tricholoma fruits under conifers. Edible, but **not** recommended (see above).

Tricholoma aurantium

Golden Tricholoma

Caps are 2–8 cm across, convex to flat, with a low knob, slippery when wet, with flattened scales and a cottony margin, and ochre to orange-brown, bruising red-brown. Flesh is white and thick, with the taste and odour of ground meal. Gills are attached, white, close, and staining rusty. Stalks are up to 6 cm tall by 1.5 cm wide, scaly, and orange-brown. Spore print is white. Widespread but not common, Golden Tricholoma fruits under conifers.

Tricholoma myomyces

Caps are 2–7 cm across, convex to flat, knobbed, dry, fibre-streaked, and grey to grey-brown or dark grey. Flesh is thin and white. Gills are attached, close, and white. Stalks are up to 6 cm tall by 4–8 mm wide and white to grey. Spore print is white. Widespread and common, this species fruits under conifers, usually late in the year after the first frost.

Tricholoma pardinum

Caps are 5–15 cm across, convex to broadly convex, becoming flat or knobbed, dry, and with the fibre-streaked surface broken up to form grey to dark grey, flat scales on a whitish ground. Gills are attached, close, and white to cream or with a pinkish tinge. Stalks are up to 12 cm tall by 4 cm wide and white or with brownish fibres. Spore print is white. Widespread but not common, this species fruits under conifers. **Poisonous.**

Tricholoma vaccinum

Caps are 3–8 cm across, convex, with a hairy margin, becoming bell-shaped, then flat, dry, and with brown to red-brown scales on a paler base. Flesh is white, staining reddish, and tastes unpleasant. Gills are attached, close, and white, staining red-brown. Stalks are up to 8 cm tall by 1.5 cm wide, fibrous to scaly, pale red-brown, and hollow. Spore print is white. Widespread and common, this species fruits under conifers.

Tricholoma virgatum

Caps are 3–8 cm across, dry, conical to bell-shaped, expanding to convex, with a sharply pointed knob (umbonate), streaked with dark fibrils, and pale grey to slate-grey or grey-brown. Gills are attached, close, and white. Stalks are up to 10 cm tall by 1.5 cm wide, white, and smooth or fibrous. Spore print is white. Widespread and common, this species fruits under conifers.

Tricholoma focale

Caps are 5–15 cm across, broadly convex, becoming flat, orange-brown, greasy, and with darker scales near the centre. Gills are white, close, and attached. Stalks are short, stout, up to 8 cm tall by 2 cm wide, off-white, and scaly. Ring is persistent. Spore print is white. Widespread but not common, this species fruits under conifers. Also known as *Armillaria focale*. Edible, but **not** recommended (see p. 298).

Tricholoma portentosum

Caps are 5–12 cm across, convex to broadly convex and knobbed, grey to grey-black, and fibre-streaked. Pleasant-tasting. Gills are attached, close, and white, sometimes tinged yellow. Stalks up to 10 cm tall by 2 cm wide, white, sometimes flushed yellowish, and fibre-streaked. Spore print is white. Widespread but not common, this species fruits under conifers in sandy locations late in the year. Edible, but **not** recommended (see p. 298).

Tricholomopsis rutilans

Caps are 4–8 cm across, convex to flat, sometimes knobbed, dry, and covered with purplish-red scales on a yellow background. Flesh is yellow and thick. Gills are attached, close, and yellow, with cottony edges. Stalks are up to 10 cm tall by 1.5 cm wide, minutely scaly, and yellow. Spore print is white. Widespread and common, this species fruits on conifer wood. Edible.

Megacollybia platyphylla

Caps are 6–20 cm across, convex, becoming flat to slightly depressed, broadly knobbed, moist to dry, powdery or smooth, margin splitting in age, and blackish-brown to grey-brown or pink-brown, streaked with darker fibrils. Flesh is thin and white and tastes unpleasant. Gills are attached, white, broad, and close to well-spaced. Stalks are up to 15 cm tall by 2 cm wide, streaked, and white. Spore print is white. Widespread and common, this species fruits on logs and stumps.

Tricholomopsis decora

Caps are 2–10 cm across, convex, becoming flat or depressed, moist to dry, hairy, and yellow to ochre, covered with tiny, dark brown scales. Flesh is thin and yellow. Gills are broadly attached, becoming short-decurrent, close, and yellow. Stalks are up to 6 cm tall by 1 cm wide, yellow, and minutely hairy to scaly. Spore print is white. Widespread but not common, this species fruits on conifer wood.

Xerula furfuracea

Caps are 3–15 cm across, broadly convex, knobbed, slippery when wet, smooth to radially wrinkled, and smoky to grey-brown or yellow-brown. Gills are attached, broad, well-spaced, and chalk-white. Stalks are up to 25 cm tall by 1.5 cm wide, brittle, smooth to powdery, brown, paler near the apex, and with a long, black, root-like extension. Spore print is white. Widespread and very common, this species fruits on or near old stumps. Edible.

Xerula megalospora

Caps are 2–8 cm across, convex, becoming flat and centrally depressed, smooth, slimy when wet, with a striate margin, and smoky-white to pale buff. Gills are attached, white, and well-spaced. Stalks are up to 13 cm tall by 1 cm wide, white, smooth, and silky to finely striate, with a root-like extension up to 8 cm long. Spore print is white. Widespread and fairly common in urban areas, this species fruits on lawns near old hardwood stumps.

The Genera *Lactarius* and *Russula*

Lactarius spp. are medium-sized to large mushrooms. They are common and widespread. There are many species, and they are often difficult to identify but a few are common and relatively distinctive and these are included in this book. *Lactarius* spp. are called "milk mushrooms" because when they are cut they **bleed** a latex-like fluid. The colour of the latex, whether or not it changes colour as it oxidizes and whether the taste is mild, sweet, bitter or acrid all help in identifying a *Lactarius*. Some *Lactarius* spp. bleed more heavily than others. Bleeding is more prolific in young specimens whereas older, drier specimens dry up and might not bleed.

Russula spp. are related to *Lactarius*. They differ in that they do not produce latex. Also, they are characteristically **brittle** and in many *Russula* spp. the gills fracture readily on handling—if you riff across the gills with your thumb, they shatter and fragment. *Russulas* are also difficult to identify, even for those knowledgeable about mushrooms. The colour of the spore print (white, cream, yellow, etc.) is important, as is the taste (pleasant, bitter, acrid, etc.) of the mushroom. As the name suggests, *Russula* spp. are often red or reddish. Unfortunately, the red pigment and other *Russula* pigments bleach rapidly with time or weathering and so the same species might come in a variety of shades. All in all, the *Russula* group is a difficult one to handle both for beginner and expert. *Russula brevipes* (p. 310) is very similar to some of the whitish *Lactarius* spp., but is easily distinguished because it doesn't bleed.

Note on Edibility:

Lactarius and Russula are large genera, each with species numbered in the hundreds. Species are often difficult to identify with certainty, so it is important to be particularly **vigilant** when deciding edibility. It is best in those genera to stick to one or a few species, eg., Delicious Lactarius (Lactarius deliciosus, p. 306). This mushroom is common, widespread and, more importantly, has several features that make it easily identifiable. As far as Russula spp. are concerned, it is difficult to tell one from the other using field characteristics and, in general, should only be eaten by the most experienced collectors and even then with **caution**.

Lactarius scrobiculatus

Caps are 5–15 cm across, convex to flat, becoming depressed or funnel-shaped, slippery when wet, hairy, becoming smooth, with a hairy margin, pale yellow to tan or ochraceous, and sometimes zonate. Flesh is white, staining yellow, and acrid. Latex is white and ageing yellow, with darker stains. Gills are broadly attached to short-decurrent, close, narrow, and white to yellow. Stalks are up to 6 cm tall by 3.5 cm wide, smooth, coloured as the cap or paler, spotted, and becoming hollow. Spore print is white to cream or yellowish. Widespread but not common, this species fruits under conifers. **Poisonous**.

Lactarius chrysorheus

Caps are 4–8 cm across, convex, becoming flat to depressed, sometimes knobbed, slippery when wet, smooth, and flesh-coloured to cinnamon or red-brown. Flesh is white and thick, and stains yellow. Latex is white, changing to yellow, and acrid. Gills are broadly attached to decurrent, close, narrow, and white to tan. Stalks are up to 8 cm tall by 1.5 cm wide, smooth or powdery, and coloured as the cap or paler. Spore print is pale yellow. Widespread and common, this species fruits on the ground under conifers. **Poisonous**.

Lactarius thyinos

Caps are 4–10 cm across, convex to flat, becoming funnel-shaped, orange to yellow-orange, zonate, and slimy to sticky. Gills are attached to decurrent, not close, and coloured as the cap. Latex is orange and doesn't stain the gills. Stalks are up to 8 cm tall by 2 cm wide, smooth, and coloured as the cap or paler. Spore print is pale yellow. Widespread but not common, this species fruits under conifers and never stains green. Edible.

Lactarius hygrophoroides

Caps are 2–7 cm across, convex to flat, becoming deeply depressed to funnel-shaped, dry, powdery or smooth, sometimes velvety, and tan-orange to orange-brown. Flesh is white and thin, with a mild taste and odour. Latex is white, unchanging, and mild. Gills are broadly attached to short-decurrent, far apart, and white to cream. Stalks are up to 5 cm tall by 1.5 cm wide and smooth. Spore print is white to cream. Widespread but not common, this species fruits under hardwoods. Edible.

Lactarius deliciosus

Delicious Lactarius

Caps are 5–12 cm across, convex, becoming flat and depressed, slippery when wet, smooth, yellow-orange to orange, blotched with blue-green, and weakly zonate. Flesh is white, staining orange or greenish. Latex is orange and mild. Gills are broadly attached to decurrent, close, narrow, and orange, staining green. Stalks are up to 7 cm tall by 2 cm wide, smooth or powdery, and coloured as the cap or paler. Spore print is cream to buff. Widespread and common, this species fruits on the ground under conifers. Edible.

Lactarius corrugis

Caps are 5–15 cm across, convex to flat and centrally depressed, rich red-brown to yellow-brown, velvety, dry, and sometimes wrinkled at the margin. Gills are attached, close, and ochre, ageing or staining brown. Stalks are up to 12 cm tall by 2.5 cm wide, dry, minutely velvety, and paler than the cap. Latex is white, abundant, and stains tissues brown. Widespread and not uncommon in our eastern regions, this species fruits under hardwood trees. Edible.

Lactarius indigo

Caps are 4–12 cm across, convex, becoming flat and depressed, smooth, indigo-blue, fading to grey, bruising green, and zonate. Flesh is blue, fading to green. Latex is dark blue and mild-tasting. Gills are attached to decurrent, close, and blue, fading to green. Stalks are up to 6 cm tall by 2.5 cm wide, smooth, and coloured as the cap or paler. Spore print is cream to yellow. Widespread but not common, this species fruits on the ground in woods. Edible.

Lactarius griseus

Caps are 1–4 cm across, convex, becoming depressed to funnel-shaped, dry, hairy, and grey to smoky-grey, with a darker centre. Flesh is white and thin. Latex is white, unchanging, and slowly acrid. Gills are broadly attached to decurrent, close to well-spaced, broad, and white to cream. Stalks are up to 6 cm tall by 5 mm wide, smooth, and grey. Spore print is white to cream. Widespread but not common, this species fruits on the ground or on very rotten wood.

Lactarius subpurpureus

Caps are 4–10 cm across, convex to flat, with a depressed centre, smooth, concentrically zonate, and vinaceous to purplish, with a striate margin. Gills are attached, not close, and coloured as the cap. Latex is scant and purplish, and tastes slightly peppery. Stalks are up to 8 cm tall by 1.5 cm wide, coloured as the cap, and smooth. Widespread and not uncommon, this species fruits under pines. Edible, but **not** recommended (see p. 304).

Lactarius lignyotus

Caps are 2–9 cm across, convex to flat, becoming depressed, often knobbed, dry, velvety, and blackish-brown to yellow-brown. Flesh is white, staining pinkish. Latex is white, turning pinkish, and mild to slightly acrid-tasting. Gills are broadly attached to decurrent, well-spaced, and white to yellowish. Stalks are up to 12 cm tall by 1.2 cm wide, velvety, and coloured as the cap. Spore print is buff. Widespread and common, this species fruits under conifers. Edible, but **not** recommended (see p. 304).

Lactarius rufus

Caps are 3–12 cm across, convex, becoming depressed to funnel-shaped, knobbed, dry, smooth, and red-brown. Flesh is white, thin, and odourless. Latex is white and unchanging, and tastes very acrid. Gills are short-decurrent, close, narrow, and ochre, becoming red-brown. Stalks are up to 11 cm tall by 1.5 cm wide, dry, smooth, powdery or hairy near the base, and coloured as the cap or paler, becoming hollow. Spore print is pinkish-buff. Widespread and common, this species fruits under pine and in sand dune areas. **Poisonous.**

Photo: Greg Thorn.

Lactarius piperatus

Caps are 5–15 cm across, convex, becoming flat and funnel-shaped, off-white to cream-coloured, and dry. Gills are attached to decurrent, crowded, and white to cream. Latex is white and tastes acrid. Stalks are up to 9 cm tall by 2.5 cm wide and coloured as the cap. Widespread and common, this species fruits under hardwood trees.

Lactarius torminosus

Caps are 5–12 cm across, convex to flat or funnel-shaped, smooth, with a hairy margin, buff to pinkish-tan, and often zonate. Flesh is white to pinkish-tan. Latex is white and unchanging, and tastes acrid. Gills are short-decurrent, close, narrow, and white to cream, tinged with pink. Stalks are up to 7 cm tall by 3 cm wide, flesh-coloured, and smooth. Spore print is cream. Widespread and common, this species fruits on the ground in woods. **Poisonous**.

Russula brevipes

Caps are 8–15 cm across, convex and umbilicate, becoming flat to funnel-shaped, dry, smooth, and white or stained brown. Flesh is white, not staining, and tastes slightly bitter. Gills are broadly attached to short-decurrent, close, and white. Stalks are up to 5 cm tall by 2 cm wide, white, and smooth to slightly hairy. Spore print is white. Widespread and common, this species fruits on the ground in woods, solitary or in groups. At first glance, this species looks like a *Lactarius*, but it does not bleed when cut.

Russula densifolia

Caps are 5–12 cm across, convex to umbilicate, becoming depressed, moist to dry, smooth, and dull white to smoky-brown. Flesh is greyish, staining red, then black, and becomes slowly bitter in taste. Gills are broadly attached to short-decurrent, close to well separated, and white to grey, staining as the flesh. Stalks are up to 6 cm tall by 3 cm wide, smooth, and white, staining as the flesh. Spore print is white. Widespread and common, this species fruits on the ground in conifer woods. Edible, but **not** recommended (see p. 304).

Russula olivacea

Caps are 10–20 cm across, convex, becoming flat, with a central depression, and olive to olivaceous, becoming red to purple-brown. Gills are attached, well-separated, and cream, becoming ochre. Stalks are up to 10 cm tall by 3.5 cm wide and white, tinged with pink. Taste is pleasant. Spore print is yellow-ochre. Widespread and not uncommon, this species fruits under hardwoods, especially beech and oak. Edible.

Russula nigricans

Blackening Russula

Caps are 7–15 cm across, convex and umbilicate, becoming flat, then funnel-shaped, smooth, and smoky-brown to brownish-black. Flesh is white, staining red, then black. Gills are attached, well-spaced, thick, white, and staining as the flesh. Stalks are up to 6 cm tall by 3 cm wide, smooth, and white, becoming smoky-brown and staining reddish. Spore print is white. Widespread and common, Blackening Russula fruits on the ground in woods. Edible, but **not** recommended (see p. 304).

Russula claroflava

Yellow Swamp Russula

Caps are 4–10 cm across, convex, becoming flat to slightly depressed, yellow to golden-yellow, slippery when wet, and smooth. Flesh is white, slowly blackening. Taste is mild. Gills are almost free and yellow to ochre, bruising black. Stalks are up to 8 cm tall by 2 cm wide, white, and smooth. Spore print is ochre. Widespread and fairly common, Yellow Swamp Russula fruits in wet spots, especially under birch or in mixed woods. Edible.

Russula aeruginea

Caps are 5–15 cm across, convex to broadly convex or flat and slightly depressed, slippery when wet, powdery to velvety, and grass-green to dull green or olive, often with a dark brown centre. Flesh is white or tinged green. Gills are attached to nearly free, close, and white to cream. Stalks are up to 8 cm tall by 2 cm wide, white to yellowish, and smooth. Spore print is creamy white. Widespread and common, this species fruits on the ground in woods. Edible.

Russula crustosa

Caps are 5–15 cm across, convex to flat, with a depressed centre and a blue-green to yellow-green base, fading to white, and covered with a mosaic of darker olivaceous patches. Gills are attached, white, and close. Stalks are up to 10 cm tall by 3 cm wide and white. Taste is mild. Spore print is ochre. Widespread but not common, this distinctive species fruits in mixed woods. Edible.

Russula fragilis

Caps are 2–5 cm across, convex to flat or slightly depressed, thin, slippery when wet, smooth, purplish to rosy, fading to white, and with a roughened, striate margin. Flesh is white and bitter-tasting. Gills are attached, close, and white. Stalks are up to 5 cm tall by 1 cm wide, white, and smooth. Widespread and fairly common, this species fruits on the ground in woods.

Russula decolorans

Caps are 5–12 cm across, convex to flat and slightly depressed, moist to dry, smooth, and salmon-pink to orange-red, with an ochre centre. Flesh is white, becoming grey. Gills are white to yellow or ochre, drying grey. Stalks are up to 12 cm tall by 2.5 cm wide, smooth or wrinkled, and white, staining grey. Spore print is ochre. Widespread and fairly common, this species fruits under conifers. Edible.

Photo: Greg Thorn.

Russula sanguinea

Rosy Russula

Caps are 4–10 cm across, convex, becoming flat and depressed, and bright rose-red, fading in age. Gills are attached, white to cream, and close. Stalks are up to 10 cm tall by 3 cm wide and white, tinged pink or red. Taste is acrid. Spore print is ochre. Widespread and fairly common, this species fruits on the ground under conifers or in mixed woods.

Russula paludosa

Caps are 6–15 cm across, convex to flat, becoming depressed, bright orange-red, fading to orange, smooth, and slimy when wet. Gills are attached, not close, and white, becoming ochre. Taste is mild to faintly bitter. Stalks are up to 10 cm tall by 3 cm wide, white, and smooth. Widespread and not uncommon, this species fruits under conifers, often in moss in wet spots. Edible.

Russula emetica

Caps are 5–10 cm across, convex to flat or slightly depressed, slippery when wet, shining when dry, smooth, and rosy- to blood-red, fading to white. The margin is roughened and striate. Flesh is white and bitter-tasting. Gills are nearly free and white. Stalks are up to 7 cm tall by 2 cm wide, smooth, and white. Spore print is white. Widespread and common, this species fruits on the ground or rotten wood. **Poisonous**.

Russula xerampelina

Caps are 5–14 cm across, convex to flat or slightly depressed, dry, smooth to powdery, and purple to olive-green to purple-red or blackish-red near the margin. Flesh is white, with a fishy smell. Gills are attached and white to cream. Stalks are up to 8 cm tall by 2.5 cm wide, smooth or powdery, and white to red, staining ochre to olivaceous. Spore print is yellow to pale ochre. Widespread and common, this species fruits on the ground in woods. Also known as *R. atropurpurea*. Edible.

Russula ventricosipes

Sand Russula

Caps are 5–12 cm across, convex to flat and depressed, sticky, brown to orange-brown, with reddish tinges, and smooth to scaly, especially at the margin. Stalks are stubby, up to 10 cm tall by 4 cm wide, whitish, and streaked or stained reddish-brown. Spore print is pale orange-yellow. Taste is unpleasant. Sand Russula fruits under pines in sandy soil around the Great Lakes and in coastal regions.

Photo: Greg Thorn.

Russula laurocerasi

Caps are up to 15 cm across, convex to broadly convex, with a depressed centre, dull yellow-ochre, and slimy when wet, with a distinctly grooved margin. Taste is unpleasant. Gills are attached, close, and white to yellowish. Stalks are up to 10 cm tall by 2.5 cm wide and dirty white to pale ochre, ageing or staining brownish. Spore print is yellow. Widespread and not uncommon, this species fruits under hardwoods.

Russula mariae

Caps are 3–9 cm across, convex to flat or depressed, dry, powdery to velvety, deep purple-red, fading to violet, and mild to slightly acrid. Gills are attached, close, narrow, and white to cream. Stalks are up to 8 cm tall by 1.5 cm wide, powdery, and pink to purple. Spore print is white to cream. Widespread and common, this species fruits on the ground in woods. Edible.

MUSHROOMS AS FOOD

Collecting wild mushrooms as edibles is a popular pastime. However, many people are armed with more enthusiasm than knowledge, so there is always a risk of poisoning from mistakes in identification. While the risk of poisoning is small, the penalty for error can be high. It pays, therefore, to err on the side of **caution**. In the beginning, it is best to stick to a few well-known and easily recognized species and then add gradually to this list. With relatively little effort, you can become fairly confident about identifications for the more common edible species.

Edible Fungi

The following (on two pages) is a list of fungi with attributes that make them desirable as edibles: they are easy to recognize; there is little chance of confusing their identity with other, possibly poisonous, species; they have been eaten by many people for many years with no ill effects; and they are widespread in their distribution and availability. There are many books available that describe methods of preparing and cooking mushrooms.

Black Morel
Morchella elata
p. 72

Yellow Morel
Morchella esculenta
p. 72

Pear-Shaped Puffball
Lycoperdon pyriforme
p. 90

Giant Puffball
Calvatia gigantea
p. 93

Comb Tooth
Hericium coralloides
p. 122

Hedgehog Mushroom
Hydnum repandum
p. 123

Chicken of the Woods
Laetiporus sulphureus
p. 149

Old Man of the Woods
Strobilomyces strobilaceus, p. 159

King Bolete
Boletus edulis
p. 163

Edible Fungi

Orange Bolete
Leccinum aurantiacum
p. 166

Slippery Jack
Suillus luteus
p. 170

Granular-Dotted Bolete
Suillus granulatus
p. 174

Horn of Plenty
Craterellus fallax
p. 190

Sidewalk Mushroom
Agaricus bitorquis
p. 192

Meadow Mushroom
Agaricus campestris
p. 193

Shaggy Mane
Coprinus comatus
p. 195

Tippler's Bane
Coprinus atramentarius
p. 196

The Gypsy
Rozites caperata
p. 219

Honey Mushroom
Armillaria mellea
p. 246

Pig's Ear
Gomphus clavatus
p. 249

Chanterelle
Cantharellus cibarius
p. 250

Fairy Ring Fungus
Marasmius oreades
p. 262

Blewit
Lepista nuda
p. 280

Oyster Mushroom
Pleurotus ostreatus
p. 291

Delicious Lactarius
Lactarius deliciosus
p. 306

Mushroom Poisoning

In general, if you experience unpleasant symptoms within an hour or so of ingesting mushrooms, then you don't usually have to worry too much about your premature demise. When the symptoms come quickly, it is usually an alimentary toxin that causes them. These symptoms will be unpleasant for a little while (nausea, vomiting, stomach cramps, diarrhea), but usually recovery is rapid and complete within a few days. If, however, you experience the first symptoms from **4** to **24** hours or more after ingestion, then you are in serious trouble and should seek medical help **immediately**. Some of the deadliest toxins do not cause any initial alimentary symptoms, but will subsequently cause serious and permanent damage to vital organs, including the kidney and liver. The effect of internal organ damage is slow to reveal itself but results in serious medical **problems** and even **death**. Don't take chances. **Any** unpleasant symptoms following mushroom ingestion should be attended to immediately.

Deadly Galerina (*Galerina autumnalis*)

Common Toxins

Amatotoxins (Amanitin and Phalloidin): These toxins are responsible for more deaths in Canada than any other toxin. They are found in Destroying Angel (*Amanita virosa*, p. 241), Death Cap (*Amanita phalloides*, not in this book), Deadly Galerina (*Galerina autumnalis*, p. 222) and a few other species. These toxins are cyclopeptides that can **destroy liver** and **kidney** functions. Usually, no symptoms are experienced for 8 to 12 hours after ingestion. General symptoms of *Amanita* poisoning include severe abdominal pains, diarrhea, retching and fever over a period of days. These symptoms are followed by a quiescent period of several more days followed by the onset of more severe symptoms, including convulsions, neuroparalysis, liver degeneration and kidney failure. The survival rate from this group of toxins is less than 50 percent. Not all *Amanita* species contain these deadly toxins and some are considered edible and of excellent flavour. We cannot recommend any *Amanita* species as edible, however, because the risks outweigh the gains. Thus, in the descriptions we have recommended **avoiding** all species of *Amanita*, even when known to be edible.

Gyromitrin: This toxin is produced by some strains of False Morel (*Gyromitra esculenta*, p. 73). Many people eat this fungus with no ill effects. In Europe, however, **hundreds** of people have been killed following ingestion of this fungus, which has now been banned as an item for sale at farmer's markets in most countries. Effects are

False Morel (*Gyromitra esculenta*)

delayed and the first symptoms occur four to six hours after ingestion. Victims will get stomach cramps, vomiting, watery diarrhea, lassitude and headaches. More severe symptoms result in loss of balance, jaundice, convulsions and coma followed by **death** in some cases. Many people eat and enjoy False Morel with no adverse effects. The same individuals, however, might consume the fungus at a later date and succumb to the toxin.

For most toxins, there is a direct dose/response relationship such that the greater the amount of toxin consumed, the more severe the symptoms experienced. Gyromitrin is an exception to this rule. No symptoms are experienced below a threshold value. Once this threshold is exceeded, then severe reactions result. The threshold level varies with individuals, so that one individual might not be affected and another dies. The toxin level will also vary with different strains, and it is claimed that some strains contain no toxin. Again, dose is based on body weight. Thus, if two people eat the same amount, the heavier person is getting a lower dose. The situation is complicated by the fact that the toxin volatilizes off at 87.5° C and thus the method of preparation will influence the toxic effects. False Morel should **not** be eaten.

Cortinarius traganus

Orellanine: This toxin is found in *Cortinarius orellanus*, one of the web-veil fungi. After consuming this poisonous mushroom, there may be no symptoms for several days or even weeks. The first symptoms are frequent urination and an intense thirst, with a burning sensation in the mouth. Later, victims experience common symptoms of mushroom poisoning, including muscle pains, nausea and headaches. Patients then suffer kidney failure and can lapse into a coma. As little as 100 g fresh weight of this mushroom can cause kidney failure. There are many hundreds of species of *Cortinarius*, and they are difficult to distinguish one from the other. Because of the dangerous nature of orellanine, **none** of the web-caps should be eaten by inexperienced collectors.

Coprine: Tippler's Bane (*Coprinus atramentarius*, p. 196), a member of the inky cap group, is a good edible. However, it must **never** be consumed with alcohol. This mushroom contains a compound that is similar to Antabuse in its effects. Antabuse is the chemical given to alcoholics to dissuade them from drinking. Drinking alcohol with *C. atramentarius* results in unpleasant symptoms, including dilation of the blood vessels, which causes flushing over the face and upper body. It takes up to three days for the toxin to be excreted from the body and alcohol must not be consumed during this period. While this toxin is not life threatening for healthy individuals, it is unpleasant for most and could cause serious problems for some.

Tippler's Bane (*Coprinus atramentarius*)

Muscarine: Muscarine was first discovered in Fly Agaric (*Amanita muscaria*, p. 239), but was later found to be at much higher concentrations (100X) in some of the little brown mushrooms belonging to the genus *Inocybe*. If muscarine-containing mushrooms are ingested, the effects are noticed in 30 minutes to two hours. The victim experiences excessive activity in the salivary glands, sweat glands and tear ducts and might become asthmatic. The symptoms are controlled by atropine.

Fly Agaric (*Amanita muscaria*)

Botrytis cinerea Grey Mould of Strawberries

Botrytis cinerea, which causes grey mould of strawberries, attacks old or damaged fruits. It produces spores as a grey-blue, powdery dust over the surface of the affected berries (or fruit). *Botrytis* spp. are a common fungus in the garden and attack a wide variety of dead, dying or delicate plant materials. This species is particularly fond of flowers and can either destroy them while still in the bud or cause spotting and loss of pigment in open flowers during wet periods.

Liberty Cap (*Psilocybe semilanceata*)

Hallucinogenic Compounds: There a number of hallucinogenic compounds produced in mushrooms that affect people consuming the fruitbodies. Muscimole and ibotenic acid are produced in Fly Agaric (*Amanita muscaria*, p. 239), and psilocybin and psilocin are produced by certain of species of *Psilocybe* (p. 204), *Panaeolina* (p. 204) and *Stropharia* (pp. 206–07). Fungi containing hallucinogenic compounds are rated as **poisonous**. The hallucinogenic *Psilocybe* species (magic mushrooms) are not common in the Great Lakes region but are common in the Maritime provinces, especially Nova Scotia and Prince Edward Island, and other coastal regions in northeastern North America.

Most hallucinogenic fungi are characterized as LBMs (Little Brown Mushrooms). There are hundreds of species of little brown mushrooms that are difficult to identify. A few LBMs contain the lethal toxins amanitin and phalloidin. Therefore, all LBMs should be **avoided**.

Haymaker's Mushroom (*Panaeolina foenisecii*)

Alimentary Toxins: There are many fungi that have never been tested for edibility. The risks are too high for casual experimentation. Often, however, people eat the wrong fungus by accident and can suffer **alimentary toxicosis.** This can involve any or all of the following symptoms: nausea, headaches, stomach cramps, vomiting and diarrhea. There are probably many fungi that produce this type of toxin and different alimentary toxins might be operative. *Russula emetica* (p. 314) is a good example of this group and is certainly well-named. In almost all cases, victims of alimentary toxins make a rapid and permanent recovery.

Penicillium Blue Mould of Foodstuffs

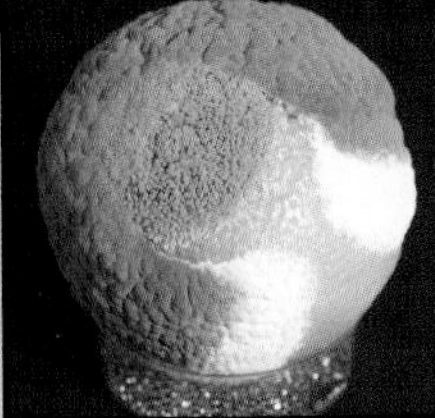

Bruised oranges develop a soft, mushy rot that is caused by *Penicillium*. If the orange is in a box or plastic bag, the fungus will break through the skin and produce a powdery, olive-green (*P. digitatum*) or blue-green (*P. italicum*) growth over the surface. The colour is because of the prolific production of the spores of the fungus. *Penicillium* is a cold temperature mould and grows happily on foodstuffs, such as cheese, bread, fruit, etc., inside the refrigerator. It is a good indicator that the food has been stored too long. In some cases *Penicillium* is deliberately introduced into foodstuffs as in the case of blue-cheese (above) where the fungus (*P. roquefortii*) is used to ripen the cheese and impart the tangy taste. The "blue" in the blue-cheese is the spore-masses!

Allergic Reactions: Finally, there is a special situation where some people develop symptoms of poisoning after ingestion of mushrooms that do no harm to others. Apparently, some people develop either allergic toxicosis or certain mushrooms contain chemicals that are toxic to some individuals but not to others. It is good sense, therefore, to eat only a **small quantity** of a mushroom that is reputed to be edible but which you have never consumed before.

Alternaria alternata

Alternaria alternata has caused grief for many people as one of the major airborne, allergy-causing fungi. It grows as a black stain on old plant material. In late fall its spores turn the leaves of corn plants black. (The photograph on the right gives an ant's eye view of this growth.) When the corn is harvested, the spores fly off in a dark cloud to cause allergy problems down-wind. *Alternaria alternata* occasionally grows inside homes on cellulose material that has stayed wet for a long time.

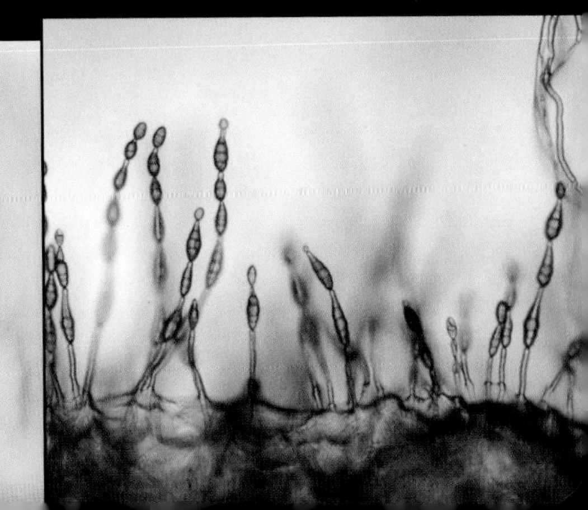

Destroying Angel
(*Amanita virosa*)

The Bastien Method for Treatment of Poisoning

In 1957 a Dr. Bastien in France developed a method to treat mushroom poisoning caused by *Amanita* species (see amatotoxins). His treatment recommended giving intravenous injections of Vitamin C (1g ascorbic acid) as soon as possible, repeated morning and evening. Also, three times a day, two gelules of nifuroxazide and two tablets of dihydrostreptomycin are taken orally (details of the doses in mg were not given). Dr. Bastien used this treatment on 15 individuals who ate Death Cap (*Amanita phalloides*) and all of them recovered.

The Lichens

Last but by no means least, I should mention the lichens. A lichen is a mutualistic symbiotic association between a species of fungus and a species of green alga. Because there are tens of thousands of species of lichens, I can only mention them here; only a book of their own would do them justice. Lichens are a biological enigma: on the one hand, they are the hardiest of "plants" and happily survive arctic cold (-50° C) and desert heat (+50° C); on the other hand, they are so sensitive to pollution that many industrial regions of the world have become devoid of lichens and are called "lichen deserts." The closer we go to the centre of an industrial city, the fewer species of lichens we can find. Good news is on the horizon: clean air legislation in many parts of the world has reduced industrial pollution to the point where the lichens are making a comeback. City dwellers should welcome their return as biological indicators of clean air.

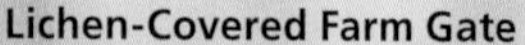

Lichen-Covered Farm Gate
In heavy shade or under exposed conditions of alternating wet and dry periods, lichens proliferate by slow, inexorable growth. Here (in photo), we see an old farm gate that has become overrun with an assortment of lichen species.

GLOSSARY

adpressed: flattened to the surface, e.g., the hairs on the cap of *Plicaturopsis crispa* (p. 147).

aethalium: large, cake-like fruitbody in slime moulds, e.g., the fruitbody of *Fuligo septica* (p. 36).

amyloid: staining grey to blue-black in Melzer's reagent, e.g., spore-mass of Destroying Angel (*Amanita virosa*, p. 241).

apothecium: fruitbody type in the sac fungi, where the hymenium is exposed at maturity.

appendiculate: when remnants of the inner veil stick to the margin, or edge, of the cap, e.g., as in *Psathyrella candolleana* (p. 203).

areolate: covered with a network of fissures, e.g., the cap of *Scleroderma areolatum* (p. 92).

ascoma: term given to any fruitbody containing asci, e.g., all sac fungi.

Ascomycota: division of the fungi containing sac fungi.

ascospore: spore produced in the sac-like ascus of the Ascomycota.

ascus: spore mother cell in the sac fungi (Ascomycota).

attached: describes gills that are connected to the stalk in mushrooms, e.g., *Tricholoma* (pp. 298–301).

autolyse: self-digest.

azonate: fruitbody surface is not in distinct zones.

Basidiomycota: division of the fungi containing those fungi in which the spore mother cell is a basidium.

basidiospore: spore produced on the spore mother cell (basidium) of Basidiomycota.

basidium: spore mother cell in the Division Basidiomycota.

calyculus: shallow, membranous cup found in some slime mould sporangia, e.g., *Hemitrichia calyculata* (p. 41).

capillitium: thread-like filaments in slime moulds that support the spore-mass and sometimes aid in spore dispersal, e.g., *Hemitrichia clavata* (p. 41).

cespitose: a cluster of fruitbodies arising close together.

chambered: covered with deep pits to give a honeycombed appearance, e.g., the cap of Yellow Morel (*Morchella esculenta*, p. 72).

clavate: club-shaped, e.g., the fruitbodies of Pestle-Shaped Coral (*Clavariadelphus pistillaris*, p. 113).

cleistothecium: tiny, spherical fruitbody type found in the sac fungi (Ascomycota). Usually >1 mm.

collection: one or several specimens of a fungus from a single site.

conidium: asexual spore, e.g., *Nyctalis asterophora*, p. 260.

context: internal tissue of a fungal fruitbody.

convex: having a shallow curve, shaped like a convex lens, e.g., the caps of many *Amanita* species (pp. 235–41).

cristate: crested like a cockscomb, e.g., the branch tips of Cockscomb Coral (*Clavulina cristata*, p. 114).

decorticated=debarked: describes a log that has lost its bark, and large sections of bare wood are exposed.

decurrent: gill attachment where the gills run down the stalk, e.g., The Miller (*Clitopilus prunulus*, p. 189).

depressed: when the central portion of the mushroom cap is sunken below the level of the edge, e.g., many *Russula* species (pp. 310–15).

disc-shaped=discoid: describes fruitbodies of sac fungi that are almost flat to saucer-shaped, e.g., fruitbodies of *Scutellinia* (p. 60).

ellipsoid: shaped like a three-dimensional ellipse=football-shaped. Often used to describe spores; sometimes used to describe the cap shape in mushrooms, e.g., young Shaggy Mane (*Coprinus comatus*, p. 195).

fibril: a fine fibre.

fibrillose: covered with delicate, fine fibres, e.g., the cap of *Crinipellis setipes* (p. 266).

fibrous: containing fibres.

flesh-coloured: pale pinkish-tan, e.g., the cap of Well's Amanita (*Amanita wellsii*, p. 235).

fleshy: imprecise term referring to the context of mushroom caps. Measured in millimetres the context is said to be thin, and measured in centimetres the context is said to be thick or fleshy.

floccose: loose, cottony to slightly woolly, often in tufts.

fluted: with deep channels, e.g., the stalk of Black Elfin Saddle (*Helvella lacunosa*, p. 75).

fruitbody: the reproductive structure of a fungus where the sexual cycle is completed and is followed by nuclear fusion, meiosis and spore production.

gelatinous: gelatin-like, able to absorb water quickly and revive, e.g., the fruitbodies of jelly fungi and some sac fungi such as *Leotia* (p. 67).

gills: the thin plates of tissue, technically called lamellae, that hang down from the underside of a mushroom cap and on which the basidia are produced.

gill spacing: based on their distance apart, gills are categorized as crowded, close or well-spaced (=well-separated). This decision is often arbitrary and is greatly influenced by the age of the fruitbody (see p. 180).

globose: globe-like, spherical, e.g., fruitbodies of *Lycoperdon pusillum* (p. 88).

hemispherical: shaped like half of a sphere, e.g., the caps of many boletes.

hoary: with a frosted appearance, e.g., the outside of the cup of *Sarcoscypha coccinea* (p. 55).

holobasidium: a club-shaped, undivided basidium; the type of basidium found in most of the Basidiomycota.

hymenium: the area where spore mother cells are compacted tightly into a fertile layer.

hypha: the vegetative or reproductive filament of a fungus; plural=hyphae.

hypothallus: cellophane-like sheath that is found below the fruitbodies of some slime moulds, e.g., *Stemonitis* (p. 43).

inner veil: also called the partial veil; a term given to the membranous sheath that protects the developing gills in mushrooms. As the cap expands, it breaks to produce a ring on the stalk or remain attached to the edge of the cap (=appendiculate).

iridescent: refracts light like a blackbird's feathers, e.g., the peridium of *Diachea leucopodia* (p. 38).

labyrinthiform: maze-like, e.g., the pores of some bracket fungi, e.g., *Daedalea* (p. 143).

lacunose: fluted, e.g., the stalks of Russell's Bolete (*Boletellus russellii*, p. 160).

latex: milky fluid exuded by species of *Lactarius* (pp. 304–09) when cut.

longitudinal: running vertically, e.g., flutes on the stalks of *Helvella lacunosa* (p. 75).

Melzer's reagent: reagent used to test for the amyloid reaction (see p. 180).

Myxomycota: division of the fungi containing the slime moulds.

mycorrhiza: mutualistic, symbiotic association between fungi and the roots of higher plants, particularly forest trees (p. 26); plural=mycorrhizae.

olivaceous: olive-green.

peristome: name for the mouth of the spore sac in the puffball group, e.g., Beaked Earthstar (*Geastrum pectinatum*, p. 95).

perithecium: flask-shaped fruitbody produced in the sac fungi (Ascomycota), usually < 1 mm tall, e.g., *Lasiosphaeria* (p. 80).

pestle-shaped: shaped like a pestle, with a spherical head and cylindrical or tapered support stalk, e.g., fruitbody of Pestle-Shaped Puffball (*Calvatia excipuliformis*, p. 94).

plasmodiocarp: the net-like fruitbody produced in a few slime moulds, e. g., *Hemitrichia serpula* (p. 41).

plasmodium: the vegetative stage of a slime mould; it appears as a disorganized mass of slime (see p. 33).

pseudoaethalium: aethalium-like fruitbody in slime moulds, where the individual sporangia can still be resolved, e.g., *Tubifera ferruginosa* (p. 46).

pubescent: downy, covered with delicate hairs, e.g., surface of *Trametes pubescens* (p. 137).

recurved: curved upwards; reflexed, e.g., the scales on the cap of Shaggy Mane (*Coprinus comatus*, p. 195).

resupinate: fruitbodies that lie flat to the surface, e.g., *Phlebia radiata* (p. 152).

reticulate: forming a raised network, e.g., the markings on the stalk of Bitter Bolete (*Tylopilus felleus*, p. 168).

ring: that part of the inner veil that remains attached to the stalk as the cap expands=annulus.

scalloped: used to describe the lobed margin of the caps of some mushrooms, e.g., *Mycena rorida* (p. 283).

sclerotium: multi-cellular fungal structures for long-term persistence over adverse periods. Sclerotia carry-over large amounts of nutrients to give the fungus a fast start in the next growing period. They are often associated with parasitic fungi, e.g., Tuberous Collybia (*Collybia tuberosa*, p. 259).

septum: name given to the wall in multi-celled fungal structures.

septate basidium: name given to the basidium in some Basidiomycota where the basidium is divided by walls into four compartments, e.g., jelly fungi (pp. 101–109).

sessile: fruitbodies that lack a stalk, e.g., *Morganella subincarnata* (p. 91).

spathulate=spatulate: shaped like a spatula; flattened at the apex and supported by a cylindrical stalk, e.g., the fruitbodies of *Spathularia* (p. 70) and *Spathulariopsis* (p. 71).

specimen: name given to a single collection of a fungus.

spines: the tooth-like projections bearing the hymenium in tooth fungi; also used to describe the stiff hairs produced on the fruitbodies of some mushrooms.

sporangium: term given to the small fruitbodies in the slime moulds that are uniform in size and shape and are produced in large numbers, e.g., *Arcyria cinerea* (p. 40).

spore: general term given to describe any reproductive unit of a fungus.

sterigma: pointed prong at the apex of the basidium on which the basidiospores are borne in the Basidiomycota.

striate: streaked or striped, e.g., the caps of many *Mycena* species (pp. 282–88).

stroma: fungal matrix in which or on which reproductive structures are produced, e.g., the finger-like "fruitbody" of *Xylaria* (pp. 79–80) or *Nectria* (p. 82).

subglobose: nearly globose, subspherical.

translucent: light shines through, almost transparent, e.g., stalks of *Mycena epipterygia* (p. 287).

truncate: with a flattened end, e.g., the fruitbody of Flat-Topped Coral (*Clavariadelphus truncatus*, p. 113).

tubes: term given to the hollow, cylindrical structures produced by boletes and bracket fungi, which are lined with spore mother cells.

tuning-fork basidium: the deeply divided, two-pronged basidium in *Dacrymyces* (p. 102). This feature is microscopic, but is used to show the relationships of *Dacrymyces* to other fungi.

umbilicate: belly–button-like; with a small, sunken hole in the middle of the cap, e.g., as *Hydnum umbilicatum* (p. 123).

umbonate: knobbed; with a distinct, raised region (broad or narrow) at the middle of the cap; found in many mushrooms, e.g., *Chroogomphus* (p. 199).

umbo: a knob (distinct, raised region) in the middle of the cap.

universal veil: term given to the membranous sac that completely encloses the young mushroom in some gill fungi. When the mushroom expands, some parts of the veil are carried up as remnants on top of the cap and/or remain at the base of the stalk as a cup, e.g., *Amanita vaginata* (p. 236).

vinaceous: wine-red.

volva: cup-like remnant of the universal veil at the base of the stalk in some mushrooms.

zonate: with well-defined zones that differ in colour and/or texture, e.g., the fruitbody of Turkey Tail (*Trametes versicolor*, p. 138).

SELECTED REFERENCES

Bessette, Alan E., Arleen R. Bessette and David W. Fischer. 1997. *Mushrooms of Northeastern North America*. Syracuse University Press, Syracuse, N.Y.

Bessette, Alan, and Walter J. Sundberg. 1987. *Mushrooms: a Quick Reference Guide to Mushrooms of North America*. Collier, New York.

Bird, C.J., and D.W. Grund. 1979. *Nova Scotian Species of* Hygrophorus. Proc. N.S. Institute of Science. 29: 1–131.

Farr, Marie. 1981. *How to Know the True Slime Moulds*. Brown, Dubuque, Iowa.

Grund, D.W., and A.K. Harrison. 1976. *Nova Scotian Boletes*. J. Cramer, Lehre, Germany.

Jenkins, D.T. 1986. Amanita *of North America*. Mad River Press, Eureka, California.

Largent, David L., and Harry D. Thiers. 1977. *How to Identify Mushrooms to Genus. II: Field Identification*. Mad River Press, Eureka, California.

Largent, David L., and Timothy J. Baroni. 1988. *How to Identify Mushrooms to Genus. VI: Modern Genera*. Mad River Press, Eureka, California.

Lincoff, G.H. 1981. *National Audubon Society Field Guide to North American Mushrooms*. Knopf, New York.

McKnight, Kent H., and Vera B. McKnight. 1987. *A Field Guide to Mushrooms of North America*. Peterson Field Guide Series. Houghton Mifflin, Boston.

Miller, Orson K. 1978. *Mushrooms of North America*. Dutton, New York.

Philips, Roger. 1991. *Mushrooms of North America*. Little and Brown, Boston.

Pomerleau, Rene. 1980. *Flore des Champignons au Quebec*. Les Editions La Presse, Montreal, Quebec.

Smith, A.H. 1947. *The North American Species of* Mycena. University of Michigan Press, Ann Arbor, Michigan.

———. 1972. *The North American Species of* Psathyrella. Memoirs of the New York Botanical Gardens #24.

Smith, A. H., Helen V. Smith and Nancy Weber. 1979. *How to Know the Gilled Mushrooms*. W. C. Brown, Dubuque, Iowa.

Smith, Helen V., and A. H. Smith. 1973. *How to Know the Non-Gilled Fleshy Fungi*. W.C. Brown, Dubuque, Iowa.

Snell, W.H., and E.A. Dick. 1970. *The Boleti of Northeastern North America*. J. Cramer, Lehre, Germany.

Stuntz, Daniel E. 1977. *How to Identify Mushrooms to Genus. IV: Keys to Families and Genera*. Mad River Press, Eureka, California.

Bold page numbers indicate primary species.

INDEX

ABOUT THE AUTHOR

George Barron graduated from the University of Glasgow in Scotland with a First Class Honours degree in Botany and later with an M.Sc. in Plant Pathology (Toronto) and a Ph.D. in Mycology (Iowa State University). From 1958 until he retired in 1993, Barron was on the faculty of the University of Guelph where he specialized on the taxonomy and biology of soil microfungi.

Barron was awarded a D.Sc. from Glasgow University in 1984 for his contributions to soil mycology. He also received the Sigma Xi distinguished research award from his colleagues at the University of Guelph in 1992 and the following year his work was recognized by the Canadian Botanical Association with the "George Lawson Medal." On the occasion of its 100th anniversary in 1996, the British Mycological Society elected a number of Centenary Fellows from around the world to recognize their mycological contributions and Barron was elected as one of this select group. At the Annual Meeting of the Mycological Society of America in Puerto Rico in 1998, Barron was honoured with the society's most prestigious award, "Distinguished Mycologist."

As a retirement project, Barron has devoted much of his time to collecting and photographing the mushrooms and other macrofungi of Canada and the northern United States.